An Observation Congruence Model

For

Systems Architecture

-- The Structure-Behavior Coalescence Approach --

William S. Chao

Structure-Behavior Coalescence

Systems Architecture = Systems Structure + Systems Behavior

The two processes are said to be observation congruent if their observational behaviors are the same, but internal behaviors may differ widely.

CONTENTS

8

10

ABOUT THE AUTHOR

Dr. William S. Chao is the CEO & founder of SBC Architecture International®. SBC (Structure-Behavior Coalescence) architecture is a systems architecture which demands the integration of systems structure and systems behavior of a system. SBC architecture applies to hardware architecture, software architecture, enterprise architecture, knowledge architecture, and thinking architecture. The core theme of SBC architecture is: Architecture = Structure + Behavior.

William S. Chao received his bachelor degree (1976) in telecommunication engineering and master degree (1981) in information engineering, both from the National Chiao-Tung University, Taiwan. From 1976 till 1983, he worked as an engineer at Chung-Hwa Telecommunication Company, Taiwan.

William S. Chao received his master degree (1985) in information science and Ph.D. degree (1988) in information science, both from the University of Alabama at Birmingham, USA. From 1988 till 1991, he worked as a computer scientist at GE Research and Development Center, Schenectady, New York, USA.

Dr. William S. Chao has been teaching at National Sun Yat-

Sen University, Taiwan since 1992 and now serves as the president of Association of Enterprise Architects, Taiwan Chapter. His research covers: systems architecture, hardware architecture, software architecture, enterprise architecture, knowledge architecture, and thinking architecture.

PART I: WHAT IS PROCESS ALGEBRA?

Algebraic Approach to the Study of Concurrent Systems

Process algebras are a diverse family of related approaches to the study of concurrent systems. Their tools are algebraic languages for the high-level description of interactions, communications, and synchronizations between a collection of independent agents or processes.

Process algebras also provide algebraic laws that allow process descriptions to be manipulated and analyzed, and permit formal reasoning about equivalences and observation congruence among processes.

Examples of Process Algebras

There are several leading algebraic approaches to modeling concurrent systems.

Communicating Sequential Processes (CSP) was first described in a 1978 paper by C. A. R. Hoare.

Arthur John Robin Gorell Milner introduced the Calculus of Communicating Systems (CCS) around 1980.

Algebra of Communicating Processes (ACP) was initially developed by Jan Bergstra and Jan Willem Klop in 1982.

Infinite-Queue SBC Process Algebra

Single-queue SBC process algebra, multi-queue SBC process algebra, and infinite-queue SBC process algebra are the three SBC process algebras.

Infinite-queue SBC process algebra evolved from CCS (Calculus of Communicating Systems).

CCS is a general process algebra language for the study of concurrent systems. Unlike CCS, infinite-queue SBC process algebra is only applicable to systems architecture.

18

PART II: MATHEMATICS OF INFINITE-QUEUE SBC PROCESSES

Interaction

An interaction represents an indivisible and instantaneous handshake or communication between two agents. The caller agent (either external environment's actor or component) communicates with the callee agent (component) through the interaction.

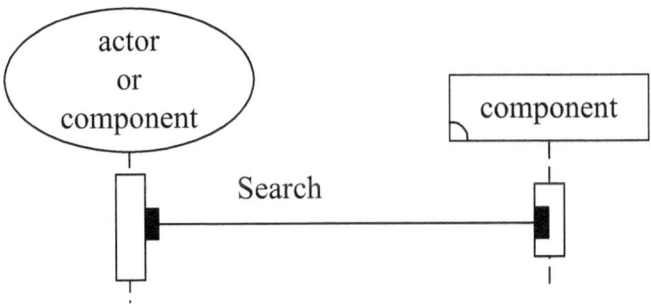

There are two ports, i.e., calling port or called port, of an interaction. The caller agent owns the "calling port" of the interaction.

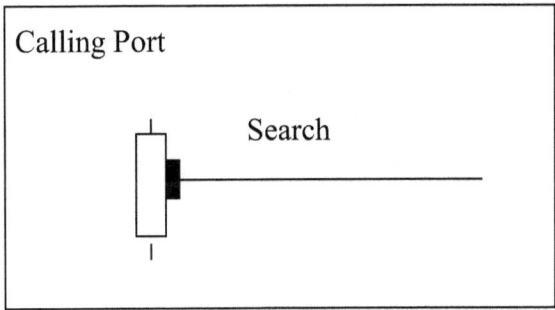

The caller agent together with the "calling port" is named the "calling action".

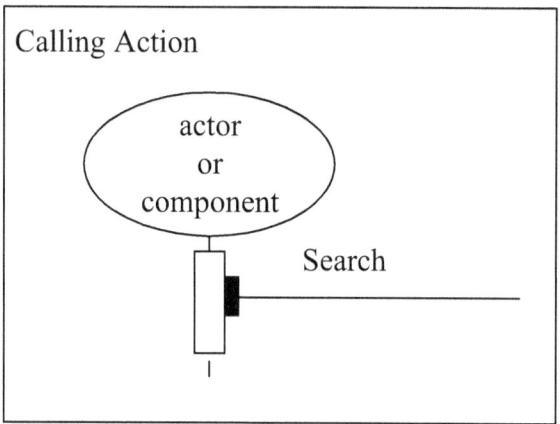

The callee agent owns the "called port" of the interaction.

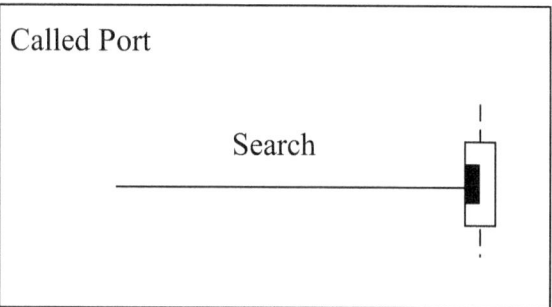

The callee agent together with the "called port" is named the "called action".

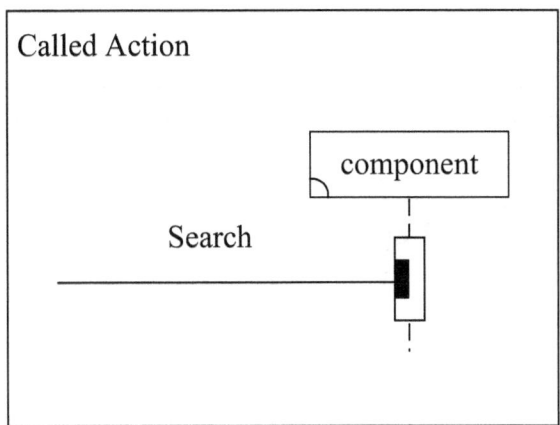

In order to simplify the interaction diagram, we will redraw it as follows.

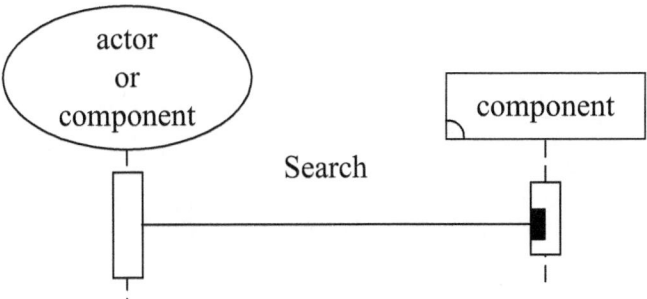

We use an internal interaction (i.e. λ) to represent their handshake or communication, if the caller agent and the callee agent are the same component.

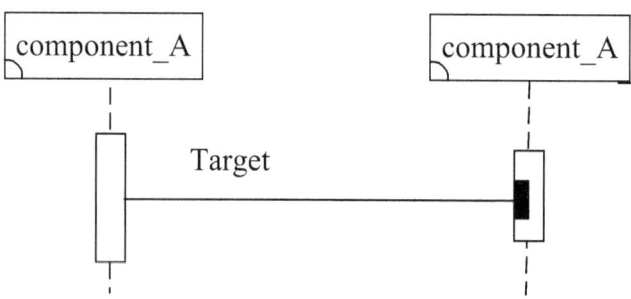

Also, we may redraw the internal interaction as follows.

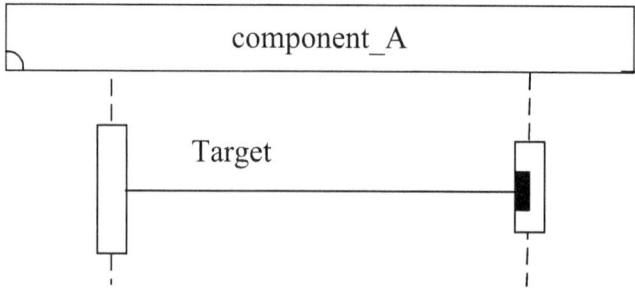

Sequentialization of Interactions

Sometimes interactions must be temporally ordered. For example, it might be desirable to specify algorithms such as: execute the "α" interaction first and then execute the "P" process later. Sequentialization of interactions can be used for such purposes.

Sequentialization of interactions, usually written as the $\alpha \bullet P$ process, indicates that it will perform the "α" interaction first and continue as the "P" process.

Parallel Composition of Processes

Parallel composition of two processes P and Q, usually written $P\|Q$, is the key primitive distinguishing the process algebras from sequential models of process executions.

Parallel composition allows the executions in P and Q to proceed simultaneously and independently.

Replication of a Process

The operators presented so far describe only finite interaction and are consequently insufficient for full computability, which includes non-terminating behavior. Replication is the operator that allows finite descriptions of infinite behavior.

For example, replication $!P$ can be understood as abbreviating the parallel composition of a countably infinite number of P processes.

Conditional Definition of a Process

A process can be defined by a one-or-more-armed conditional expression.

For example, the process (**if** $cond_1$ **then** P_1)+(**if** $cond_2$ **then** P_2)...+(**if** $cond_j$ **then** P_j) will proceed as the process P_1 if the "$cond_1$" value is true, or proceed as the process P_2 if the "$cond_2$" value is true,..., or proceed as the process P_j if the "$cond_j$" value is true.

Null Process

Process algebras generally also include a null process, denoted as *STOP*, which has no interaction points. It is utterly inactive and its sole purpose is to act as the inductive anchor on top of which more interesting processes can be generated.

The process "*STOP•P*" (i.e. sequential composition of processes *STOP* and *P*) equals to the process "*STOP*".

$$STOP \bullet P \quad = \quad STOP$$

The process "*P||STOP*" (i.e. parallel composition of processes *P* and *STOP*) equals to the process "*STOP||P*" (i.e. parallel composition of processes *STOP* and *P*) which equals to the process "*P*".

$$P \parallel STOP \quad = \quad STOP \parallel P \quad = \quad P$$

PART III: THE STRUCTURE-BEHAVIOR COALESCENCE APPROACH

Structure-Behavior Coalescence Means to Integrate the Systems Structure and Systems Behavior

Systems structure and systems behavior are the two most prominent views of a system, integrating the systems structure and systems behavior is apparently the best way to achieve an integrated whole of a system.

If we are not able to integrate the systems structure and systems behavior, then there is no way that we are able to integrate the whole system.

Structure-behavior coalescence (SBC) provides an elegant way to integrate the systems structure and systems behavior of a system. In other words, SBC facilitates an integrated whole of a system.

Core Theme of Structure-Behavior Coalescence

The core theme of structure-behavior coalescence is: "Systems Architecture = Systems Structure + Systems Behavior".

Systems Structure X + Systems Behavior X

One systems structure will draw forth one systems behavior. That is, the systems behavior is attached to or built on the systems structure in the SBC approach.

In other words, the systems behavior can not exist alone; it must be loaded on the systems structure just like a cargo is loaded on a ship.

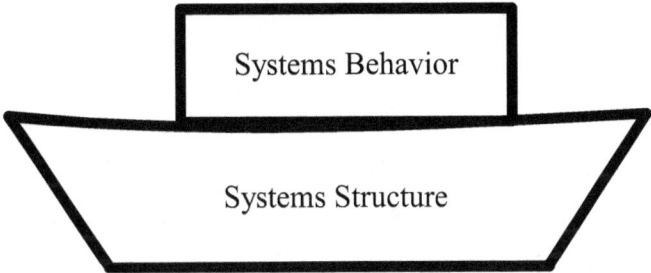

Interactions among Components and Actors to Draw Forth the Systems Behavior

In a system, if the components, and among them and the external environment's actors to interact (or handshake), these interactions will draw forth the systems behavior.

We conclude that "interaction" plays an important factor in integrating the systems structure and systems behavior for a system.

The overall behavior of a system consists of many individual behaviors. Each individual behavior represents an execution path. We use an interaction flow diagram (IFD) to demonstrate this individual behavior.

SBC View Model

In the SBC view model, dimension 1 stands for the evolution&motivation view which contains the strategy/version 1, strategy/version 2, strategy/version 3, strategy/version 4, strategy/version ∞ views; dimension 2 stands for the multi-level view which contains the concept, analysis, design, and implementation views; dimension 3 stands for the systemic view which contains the structure and behavior views.

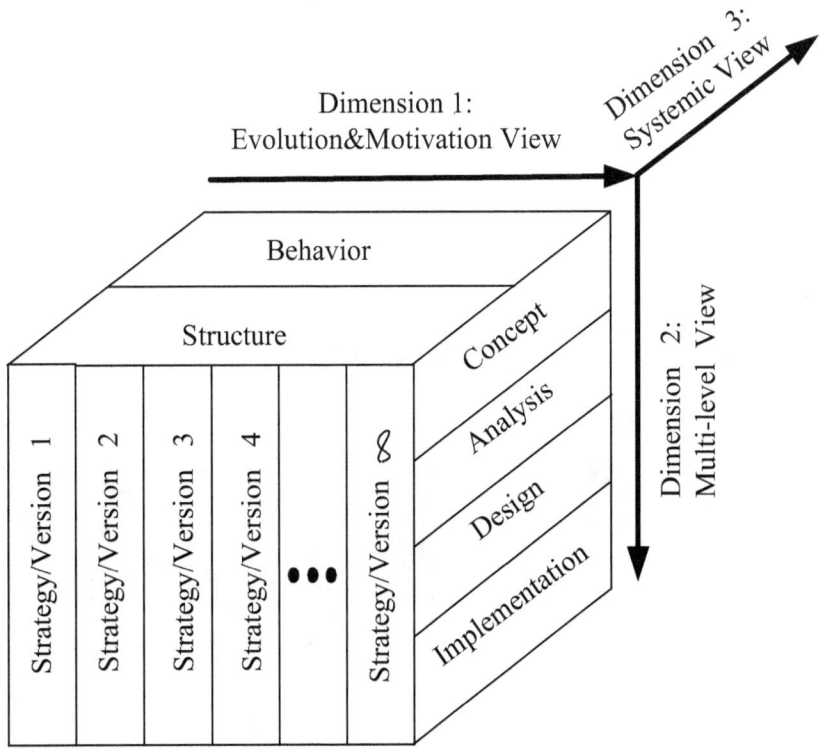

Collection of All Interaction Flow Diagrams Defines the Systemic View of Systems Architecture

The collection of all interaction flow diagrams defines the integration of systems structure and systems behavior of a system.

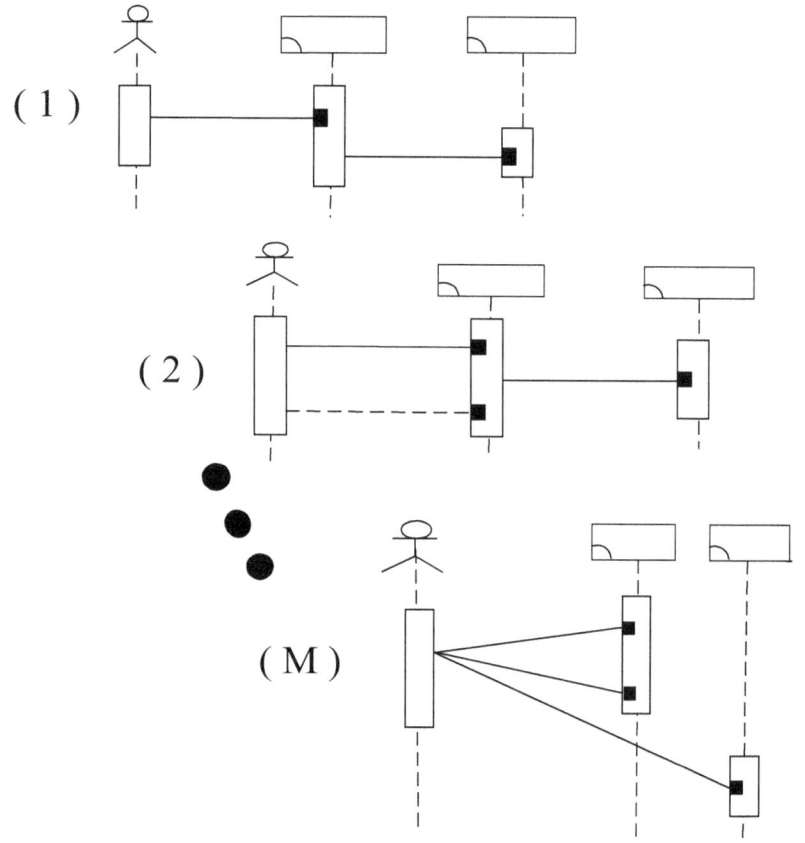

That is, the collection of all interaction flow diagrams defines the systemic view of systems architecture.

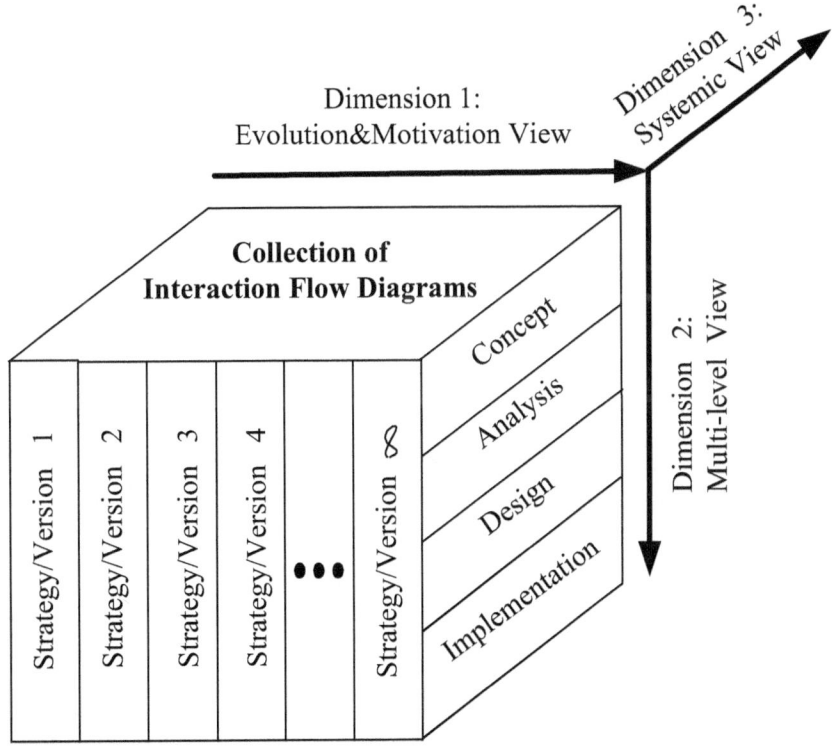

PART IV: LANGUAGE CONSTRUCTS OF INFINITE-QUEUE SBC PROCESS ALGEBRA

Backus-Naur Form of Infinite-Queue SBC Processes

The set of infinite-queue SBC (i.e. Structure-Behavior Coalescence) processes for (the structural composition of) the systemic view of each level (i.e. Concept, Analysis, Design, Implementation) is defined by the following BNF grammar:

(1) <Structural Composition of Infinite-Queue_SBC_Process_of_Each_Level's_Systemic_View> ::=
    <Infinite-Queue_SBC_Process_of_Each_Level's_Systemic_View>[*f*]

(2) <Infinite-Queue_SBC_Process_of_Each_Level's_Systemic_View> ::=
    <Parallel_of_!IFD>

(3) <Parallel_of_!IFD> ::=   *STOP*
    | <!IFD> "||" <Parallel_of_!IFD>

(4) <!IFD> ::=   <IFD>
    | <IFD> "||" <!IFD>

(5) <IFD> ::=
    <Type_1_Expression> "● " <Zero_Or_More_Expressions>

(6) <Zero_Or_More_Expressions> ::=   *STOP*
  | <Type_1_Or_2_Expression> " ● " <Zero_Or_More_Expressions>

(7) <Type_1_Or_2_Expression>   ::= <Type_1_Expression>
    | <Type_2_Expression>

(8) <Type_1_Expression>   ::= <Type_1_Interaction>
  | <Condition> <Type_1_Interaction>
    {"+" <Condition> <Type_1_Interaction>}

(9) <Type_2_Expression>   ::= <Type_2_Interaction>
  | <Condition> <Type_2_Interaction>
    {"+" <Condition> <Type_2_Interaction>}

(10) <Type_1_Interaction> ::= <Actor> <Operation_Call_Or_Return>
        <Operation_Call_Or_Return_Formula> <Component>

(11) <Type_2_Interaction> ::= <Component> <Operation_Call_Or_Return>
        <Operation_Call_Or_Return_Formula> <Component>

A Component Interacting with another Component Defines the Type_2 Interaction

Rule 11 describes that a component interacting with another component defines the type_2 interaction.

Rule 11
<Type_2_Interaction> ::= <Component> <Operation_Call_Or_Return> <Operation_Call_Or_Return_Formula> <Component>

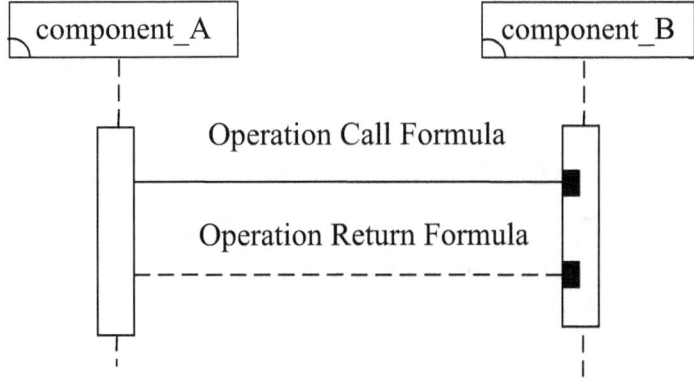

An Actor Interacting with a Component Defines the Type_1 Interaction

Rule 10 describes that an actor interacting with a component defines the type_1 interaction.

Rule 10
<Type_1_Interaction> ::= <Actor> <Operation_Call_Or_Return> <Operation_Call_Or_Return_Formula> <Component>

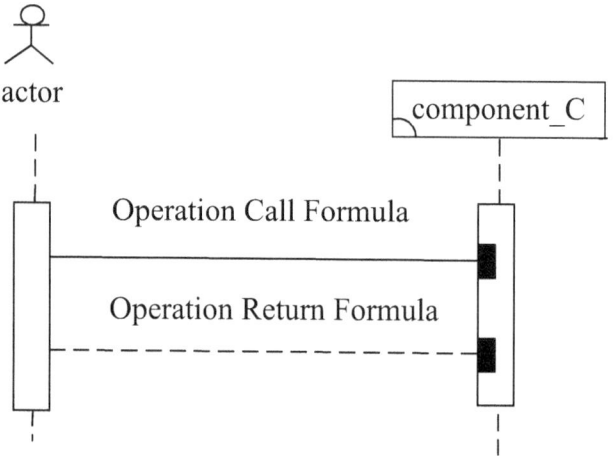

Type_2 Expression is either a Type_2 Interaction or a Conditional Type_2 Interaction

Rule 9 describes that the type_2 expression (i.e. Type_2_Expression) is either a type_2 interaction (i.e. Type_2_Interaction) or a conditional type_2 interaction (i.e. one-or-more-armed conditional expression of Type_2_Interaction).

Rule 9

<Type_2_Expression>   ::= <Type_2_Interaction>
  | <Condition> <Type_2_Interaction>
    {"+" <Condition> <Type_2_Interaction>}

Type_1 Expression is either a Type_1 Interaction or a Conditional Type_1 Interaction

Rule 8 describes that the type_1 expression (i.e. Type_1_Expression) is either a type_1 interaction (i.e. Type_1_Interaction) or a conditional type_1 interaction (i.e. one-or-more-armed conditional expression of Type_1_Interaction).

Rule 8
<Type_1_Expression>   ::= <Type_1_Interaction> | <Condition> <Type_1_Interaction>   {"+" <Condition> <Type_1_Interaction>}

Type_1_Or_2 Expression is either Type_1 or Type_2

Rule 7 describes that the type_1_or_2 expression (i.e. Type_1_Or_2_Expression) is either a type_1 expression (i.e. Type_1_Expression) or a type_2 expression (i.e. Type_2_Expression).

Rule 7
<Type_1_Or_2_Expression> ::=   <Type_1_Expression> | <Type_2_Expression>

Zero or More Expressions either are a Null Process or Consist of a Type_1_Or_2 Expression, Followed by a Sequential Composition, and Followed by Zero or More Expressions

Rule 6 describes that zero or more expressions (i.e. Zero_Or_More_Expressions) either a) are a null process (i.e. *STOP)*, or b) consist of a type_1_or_2 expression (i.e. Type_1_Or_2_Expression), followed by a sequential composition (i.e. •), and followed by zero or more expressions (i.e. Zero_Or_More_Expressions).

Rule 6

<Zero_Or_More_Expressions> ::=
 	
  STOP
 	
  | <Type_1_Or_2_Expression> "●" <Zero_Or_More_Expressions>

An Interaction Flow Diagram Consists of a Type_1 Expression, Followed by a Sequential Composition, and Followed by Zero or More Expressions

Rule 5 describes that an interaction flow diagram (i.e. IFD) consists of a type_1 expression (i.e. Type_1_Expression), followed by a sequential composition (i.e. ●), and followed by zero or more expressions (i.e. Zero_Or_More_Expressions).

Rule 5
<IFD> ::= <Type_1_Expression> "●" <Zero_Or_More_Expressions>

Parallel Composition of a Countably Infinite Number of an Interaction Flow Diagram Defines the Replicated Interaction Flow Diagram

Rule 4 describes that the parallel composition (i.e. ||) of a countably infinite number of an interaction flow diagram (i.e. IFD) defines the replicated interaction flow diagram (i.e. !IFD).

Rule 4
<!IFD> ::=   <IFD> | <IFD> "||" <!IFD>

A NULL Process or the Parallel Composition Composing All Replicated Interaction Flow Diagrams Defines the Parallel of All Replicated Interaction Flow Diagrams

Rule 3 describes that we use either a) a null process (i.e. *STOP*), or b) the parallel composition (i.e. ‖) composing all replicated interaction flow diagrams (i.e. !IFD), to define the parallel of all replicated interaction flow diagrams (i.e. Parallel_of_!IFD).

Rule 3

<Parallel_of_!IFD> ::=
            STOP
         | <!IFD>   "‖"   <Parallel_of_!IFD>

Parallel of All Replicated Interaction Flow Diagrams Defines the Infinite-Queue SBC Process of the Systemic View of Each Level

Rule 2 describes that the parallel of all replicated interaction flow diagrams (i.e. Parallel_of_!IFD), in which each interaction flow diagram may replicate itself a countably infinite times (i.e. !IFD), defines the infinite-queue SBC process of the systemic view of each level (i.e. Concept, Analysis, Design, Implementation).

Rule 2
<Infinite-Queue_SBC_Process_of_Each_Level's_Systmic_View> ::=   <Parallel_of_!IFD>

Renaming the Infinite-Queue SBC Process of Each Level's Systemic View Defines the Structural Composition of Infinite-Queue SBC Process of the Systemic View of Each Level

Rule 1 describes that applying the renaming function (i.e. $[f]$) to the infinite-queue SBC process of each level's (i.e. Concept, Analysis, Design, Implementation) systemic view defines the structural composition of infinite-queue SBC process of the systemic view of each level (i.e. Concept, Analysis, Design, Implementation).

Rule 1
<Structural Composition of Infinite-Queue_SBC_Process_of_Each_Level's_Systemic_View> ::= <Infinite-Queue_SBC_Process_of_Each_Level's_Systemic_View> $[f]$

For each renaming function f, the renaming combinatory $[f]$, postfixed to a process or an interaction, has the effect of renaming the components (of the process or interaction) as dictated by f. We often write $C'_1/C_1,\ldots,\ C'_n/C_n$ for the renaming function for which $f(C_i) = C'_i$ for i = 1,..., n.

After renaming, an interaction may be internalized to be an internal interaction (i.e. λ) if it describes that a component interacts with itself.

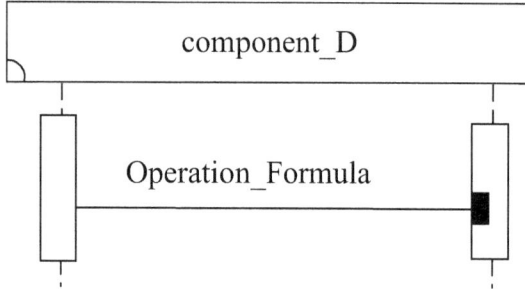

PART V: TRANSITIONAL SEMANTICS OF INFINITE-QUEUE SBC PROCESS ALGEBRA

Transitional Semantics

We assume an infinite set Δ of type_1_or_2 interactions, and use a, b to range over Δ. Henceforward we let $\Omega = \Delta \cup \{\lambda\}$, the set of all possible interactions, and use α, β to range over Ω. Further, we let Φ be the set of process Constants, and use A, B to range over Φ. We let Π be the set of processes, and use P, Q to range over Π. We let Ψ be the set of process expressions, and use E, F to range over Ψ. We let Γ be the set of components, and use C, D to range over Γ.

Entity set	Entity name	Type of entity
Δ	a, b,...	type_1_or_2 interactions
	λ	internal interaction
Ω	α, β,...	type_1_or_2 or internal interactions
Ω^*	s,...	interaction sequences
	f,...	renaming functions
Φ	A, B,...	process Constants
Π	P, Q,...	processes
Ψ	E, F,...	process expressions
Γ	C, D,...	components
	S	bisimulations

In giving meaning to the infinite-queue SBC process algebra, we shall use the following labelled transition system (LTS)

$$(\Psi, \Omega, \rightarrow)$$

which consists of a set Ψ of process expressions, a set Ω of "type_1_or_2 or internal interactions", and a transition relation $\rightarrow \subseteq \Psi \times \Omega \times \Psi$ where $(E_i, \alpha, E_j) \in \rightarrow$ is written as $E_i \xrightarrow{\alpha} E_j$.

The semantics for Ψ consists in the transition rules of each transition relation $\rightarrow$ over $\Psi \times \Omega \times \Psi$. These transition rules will follow the structure of expressions.

We give the complete set of transition rules; the names Prefix, Parallel, Rename, and Constant indicate that the rules are associated respectively with Prefix, Parallel Composition, and Structural Composition and with Constants.

Prefix $\qquad \dfrac{}{\alpha \bullet E \xrightarrow{\alpha} E}$

Parallel$_1$ $\qquad \dfrac{E \xrightarrow{\alpha} E'}{E \| F \xrightarrow{\alpha} E' \| F}$

Parallel$_2$ $\qquad \dfrac{F \xrightarrow{\alpha} F'}{E \| F \xrightarrow{\alpha} E \| F'}$

Rename $\qquad \dfrac{E \xrightarrow{\alpha} E'}{E[f] \xrightarrow{\alpha[f]} E'[f]}$

Constant $\qquad \dfrac{P \xrightarrow{\alpha} P'}{A \xrightarrow{\alpha} P'} \quad (A \stackrel{\text{def}}{=} P)$

Rule of Prefix

The rule for Prefix can be read as follows: Under any circumstances, we always infer $\alpha \bullet E \xrightarrow{\alpha} E$. That is, an expression, with an interaction prefixed to it, will use this interaction to accomplish the transition.

$$\frac{}{\alpha \bullet E \xrightarrow{\alpha} E}$$

Rules of Parallel Composition

There are two transition rules for parallel composition. Rule Parallel$_1$ indicates that from $E \xrightarrow{\alpha} E'$ we shall infer $E\|F \xrightarrow{\alpha} E'\|F$.

$$
\frac{E \xrightarrow{\alpha} E'}{E\|F \xrightarrow{\alpha} E'\|F}
$$

Rule Parallel$_2$ indicates that from $F \xrightarrow{\alpha} F'$ we shall infer $E\|F \xrightarrow{\alpha} E\|F'$.

$$
\frac{F \xrightarrow{\alpha} F'}{E\|F \xrightarrow{\alpha} E\|F'}
$$

Rule of Structural Composition

The rule for Structural Composition can be read as follows:

Rule Rename indicates that from $E \xrightarrow{\alpha} E'$ we shall infer $E[f] \xrightarrow{\alpha[f]} E'[f]$.

$$E \xrightarrow{\alpha} E'$$
$$\overline{\qquad\qquad\qquad}$$
$$E[f] \xrightarrow{\alpha[f]} E'[f]$$

Rule of Constants

The rule for Constants can be read as follows: the rule of Constants asserts that each Constant has the same transitions as its defining expression.

$$\frac{P \xrightarrow{\alpha} P'}{A \xrightarrow{\alpha} P'} \quad (A \stackrel{\text{def}}{=} P)$$

Examples of Transitional Semantics

As a first example, consider the infinite-queue SBC process Constant A is defined as $(a_1 \bullet a_2 \bullet STOP) \| (b_1 \bullet b_2 \bullet b_3 \bullet STOP)$. The following transition graph shows the semantics of process A.

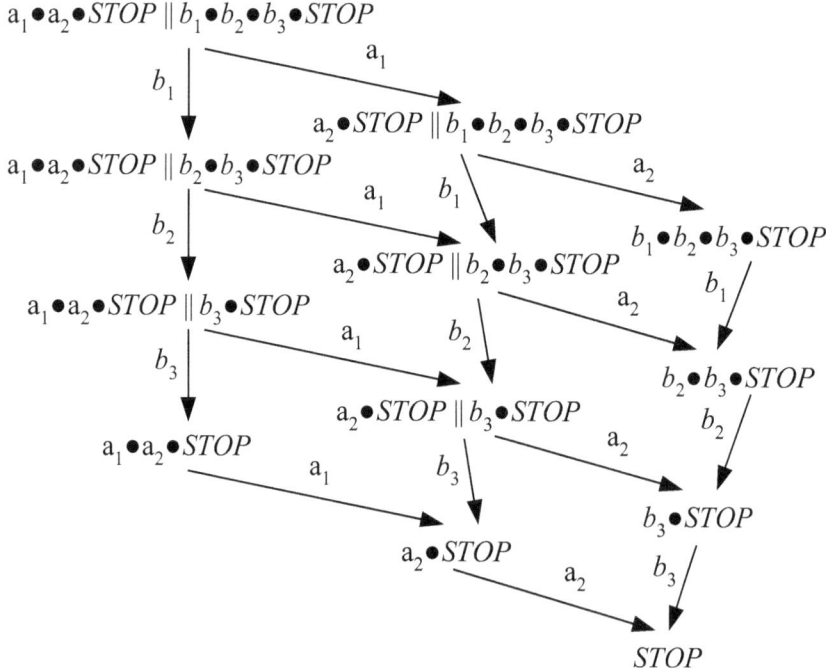

As a second example, consider the infinite-queue SBC process Constant B is defined as $(\lambda \bullet STOP) \| (a \bullet STOP) \| (b \bullet STOP)$. The following transition graph shows the semantics of process B.

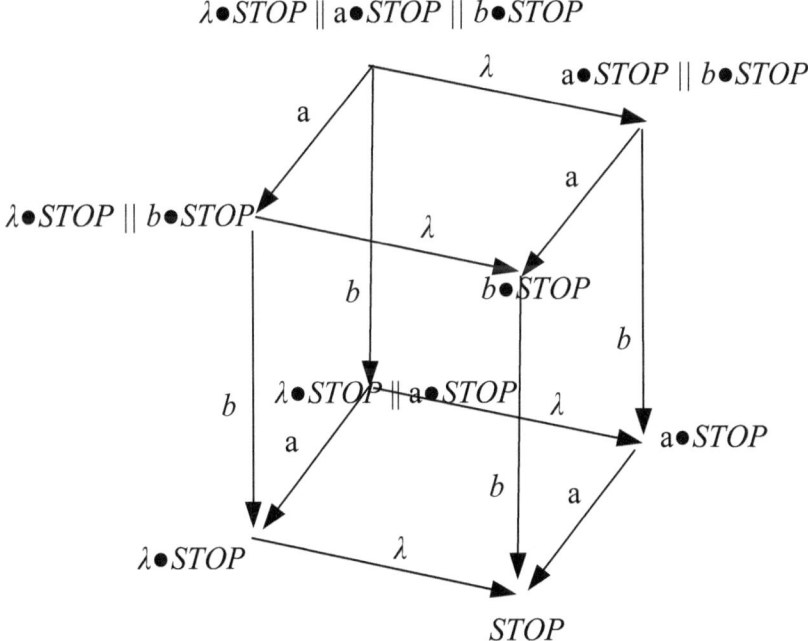

PART VI: OBSERVATION CONGRUENCE THEORY OF INFINITE-QUEUE SBC PROCESS ALGEBRA

70

Preliminary Definitions

A few preliminary definitions are needed. Definition 6.1 and Definition 6.2 are easy to comprehend.

Definition 6.1 $\quad\widehat{\alpha} = \varepsilon$ (the empty sequence), if $\alpha = \lambda$

$\qquad\qquad\qquad = \alpha$, if $\alpha \neq \lambda$

Definition 6.2 $\quad E \overset{\alpha}{\Longrightarrow} E'\quad$ iff $\quad E\ (\overset{\lambda}{\rightarrow})^*\ \overset{\alpha}{\rightarrow}\ (\overset{\lambda}{\rightarrow})^*\ E'$

$\qquad\qquad\quad E \overset{\varepsilon}{\Longrightarrow} E'\quad$ iff $\quad E\ (\overset{\lambda}{\rightarrow})^*\ (\overset{\lambda}{\rightarrow})^*\ E'$

Observation Equivalence

To achieve the observation congruence, we need to define the observation equivalence first. Definition 6.3 and Definition 6.4 together define the observation equivalence.

Definition 6.3 A binary relation $S \subseteq \Pi \times \Pi$ over multi-queue SBC processes is a *bisimulation* if $(P, Q) \in S$ implies, for all or each $\alpha \in \Omega$,

(i) Whenever $P \overset{\alpha}{\to} P'$ then, for some Q', $Q \overset{\widehat{\alpha}}{\Longrightarrow} Q'$ and $(P', Q') \in S$

(ii) Whenever $Q \overset{\alpha}{\to} Q'$ then, for some P', $P \overset{\widehat{\alpha}}{\Longrightarrow} P'$ and $(P', Q') \in S$

Definition 6.4 P and Q are observation equivalent, written $P \approx Q$, if $(P, Q) \in S$ for some bisimulation S. That is,

$$\approx = \bigcup \ (S : S \text{ is a bisimulation})$$

Observation Congruence

Once we have the definition of observation equivalence, i.e. $\approx$, we shall use it to define the observation congruence.

Definition 6.5 P and Q are observation congruent, written $P = Q$, if for all α

(i) Whenever $P \xrightarrow{\alpha} P'$ then, for some Q', $Q \xRightarrow{\alpha} Q'$ and $P' \approx Q'$;

(ii) Whenever $Q \xrightarrow{\alpha} Q'$ then, for some P', $P \xRightarrow{\alpha} P'$ and $P' \approx Q'$.

Examples of Observation Congruence Verification

As a first example, consider the infinite-queue SBC process P_1 is syntactically represented as $(a \bullet STOP) \| (b \bullet STOP)$ and the infinite-queue SBC process Q_1 is syntactically represented as $(a \bullet \lambda \bullet STOP) \| (b \bullet \lambda \bullet STOP)$, which are represented by the following transition graphs.

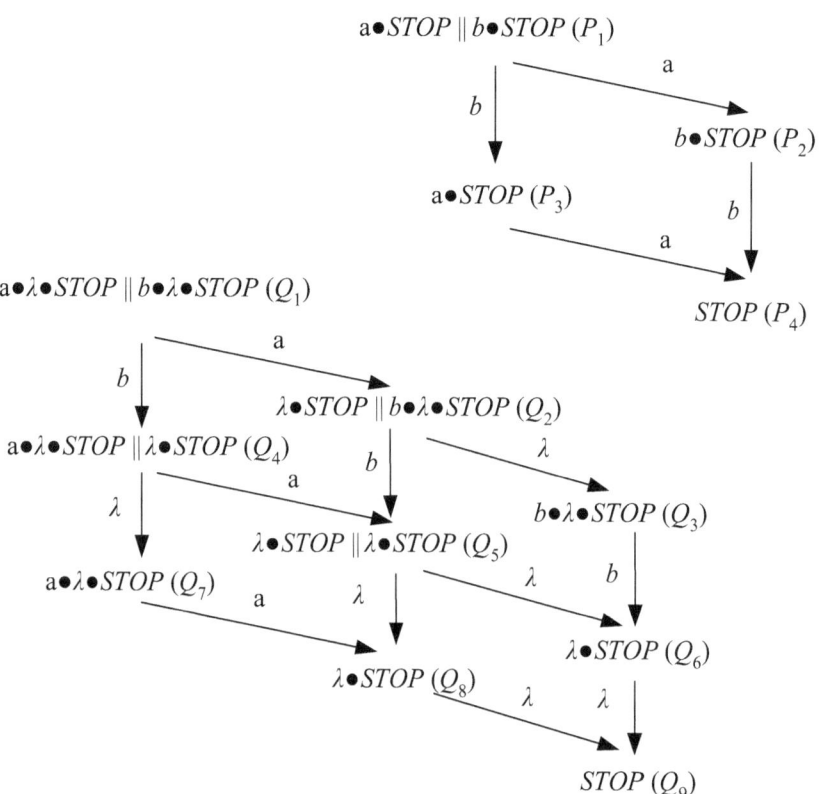

We can easily verify that $S = \{(P_1, Q_1), (P_2, Q_2), (P_2, Q_3), (P_3, Q_4), (P_3, Q_7), (P_4, Q_5), (P_4, Q_6), (P_4, Q_8), (P_4, Q_9)\}$ is a bisimulation.

Using the S bisimulation, we then are able to verify that P_1 and Q_1 are observation congruent because (1) $P_1 \overset{a}{\rightarrow} P_2$, then we have Q_2 that $Q_1 \overset{a}{\Rightarrow} Q_2$ and $P_2 \overset{\approx}{\sim} Q_2$, and (2) $P_1 \overset{b}{\rightarrow} P_3$, then we have Q_4 that $Q_1 \overset{b}{\Rightarrow} Q_4$ and $P_3 \overset{\approx}{\sim} Q_4$, and (3) $Q_1 \overset{a}{\rightarrow} Q_2$, then we have P_2 that $P_1 \overset{a}{\Rightarrow} P_2$ and $P_2 \overset{\approx}{\sim} Q_2$, and (4) $Q_1 \overset{b}{\rightarrow} Q_4$, then we have P_3 that $P_1 \overset{b}{\Rightarrow} P_3$ and $P_3 \overset{\approx}{\sim} Q_4$.

As a second example, consider the infinite-queue SBC process P_1 is syntactically represented as $b \bullet STOP$ and the infinite-queue SBC process Q_1 is syntactically represented as $\lambda \bullet b \bullet STOP$, which are represented by the following transition graphs.

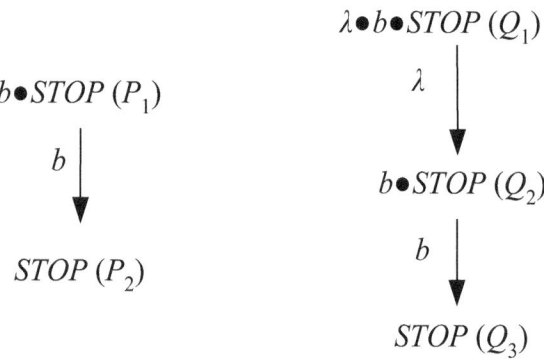

We can easily verify that $S = \{(P_1, Q_1), (P_1, Q_2), (P_2, Q_3)\}$ is a bisimulation.

Using the S bisimulation, we then are able to verify that P_1 and Q_1 are not observation congruent because for $Q_1 \overset{\lambda}{\to} Q_2$ and we find no P_m that $P_1 \overset{\lambda}{\Rightarrow} P_m$ and $P_m \overset{\approx}{\sim} Q_2$.

As a third example, consider the infinite-queue SBC process P_1 is syntactically represented as $b \bullet STOP$ and the infinite-queue SBC process Q_1 is syntactically represented as $(b \bullet STOP) \| (b \bullet STOP)$, which are represented by the following transition graphs.

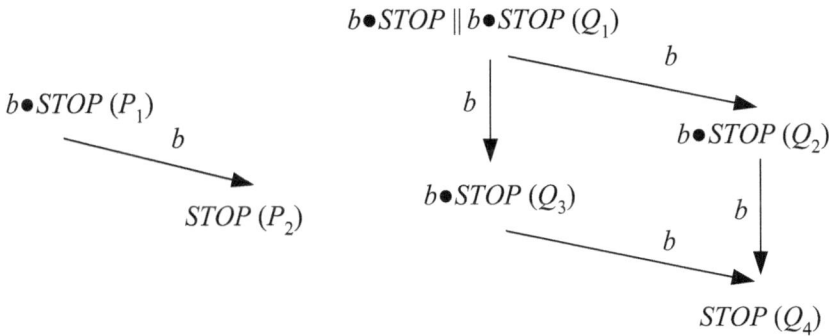

We can easily verify that $S = \{(P_1, Q_2), (P_1, Q_3), (P_2, Q_4)\}$ is a bisimulation.

Using the S bisimulation, we then are able to verify that P_1 and Q_1 are not observation congruent because for $Q_1 \overset{b}{\rightarrow} Q_2$ and we find no P_n that $P_1 \overset{b}{\Rightarrow} P_n$ and $P_n \overset{\approx}{} Q_2$.

PART VII: MULTI-LEVEL VIEW OF SYSTEMS ARCHITECTURE

SBC View Model of Systems Architecture

In the SBC view model of systems architecture, dimension 2 stands for the multi-level view which contains the concept, analysis, design, and implementation views; dimension 3 stands for the systemic view which contains the collection of all interaction flow diagrams.

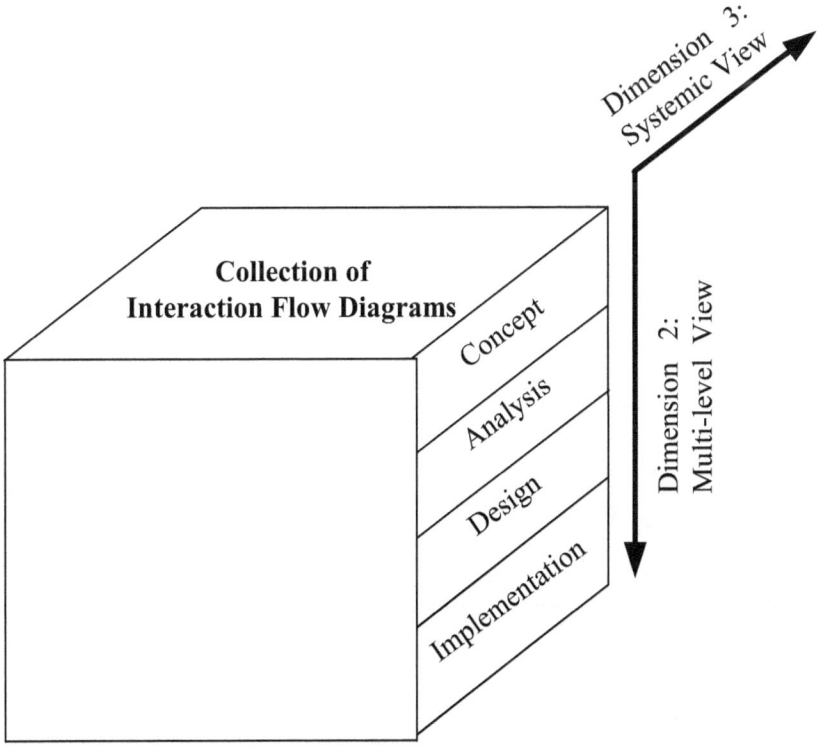

In the SBC multi-level view of systems architecture, analysis view is one level down   structural decomposition (with observation congruence verification) of the concept view; design view is one level down   structural decomposition (with observation congruence verification) of the analysis view; implementation view is one level down structural decomposition (with observation congruence verification) of the design view.

Analysis View is One Level down Structural Decomposition of the Concept View

To demonstrate that the analysis view is one level down structural decomposition (with observation congruence verification) of the concept view, we need to go through the following four steps:

(A) Get the infinite-queue SBC process of the concept view.

(B) Get the infinite-queue SBC process of the analysis view.

(C) Get the infinite-queue SBC process of the structural composition of the analysis view.

(D) Verify that there is observation congruence of "the concept view" and "the structural composition of the analysis view".

Design View is One Level down Structural Decomposition of the Analysis View

To demonstrate that the design view is one level down structural decomposition (with observation congruence verification) of the analysis view, we need to go through the following four steps:

(A) Get the infinite-queue SBC process of the analysis view.

(B) Get the infinite-queue SBC process of the design view.

(C) Get the infinite-queue SBC process of the structural composition of the design view.

(D) Verify that there is observation congruence of "the analysis view" and "the structural composition of the design view".

Implementation View is One Level down Structural Decomposition of the Design View

To demonstrate that the implementation view is one level down structural decomposition (with observation congruence verification) of the design view, we need to go through the following four steps:

(A) Get the infinite-queue SBC process of the design view.

(B) Get the infinite-queue SBC process of the implementation view.

(C) Get the infinite-queue SBC process of the structural composition of the implementation view.

(D) Verify that there is observation congruence of "the design view" and "the structural composition of the implementation view".

PART VIII: CASE STUDY --
KURDI UNIVERSITY

SBC View Model of Kurdi University

In the SBC view model of *Kurdi University*, dimension 2 stands for the multi-level view which contains the concept, analysis, design, and implementation views; dimension 3 stands for the systemic view which contains the collection of all interaction flow diagrams.

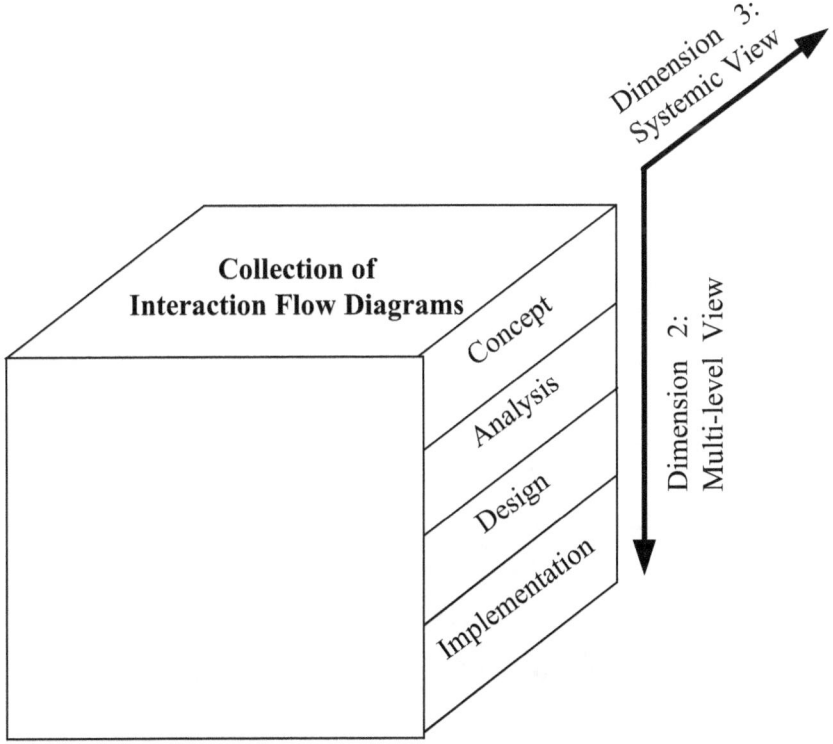

In the SBC multi-level view of *Kurdi University*, analysis view is one level down structural decomposition (with observation congruence verification) of the concept view; design view is one level down structural decomposition (with observation congruence verification) of the analysis view; implementation view is one level down structural decomposition (with observation congruence verification) of the design view.

Infinite-Queue SBC Process of the Concept View

We draw the Architecture Hierarchy Diagram (AHD) of the concept view of *Kurdi University* as follows:

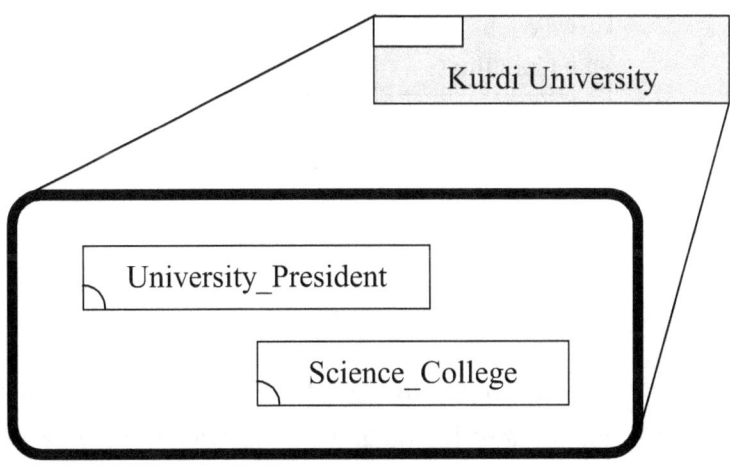

The overall behavior of the concept view of *Kurdi University* includes two behaviors: *Study_Calculus_Course* and *Study_Algebra_Course*. Each of them is described by an individual IFD.

An IFD of the *Study_Calculus_Course* behavior is shown below. First, actor *Student* interacts with the *University_President* component through the *University_Teach_Calculus* operation call interaction. Next, component *University_President* interacts with the *Science_College* component through the *College_Teach_Calculus* operation call interaction.

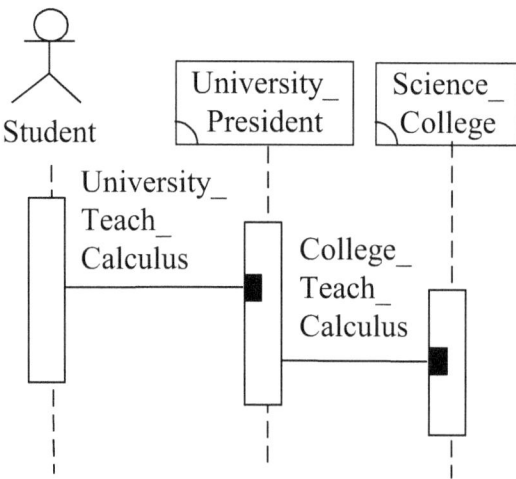

An IFD of the *Study_Algebra_Course* behavior is shown below. First, actor *Student* interacts with the *University_President* component through the *University_Teach_Algebra* operation call interaction. Next, component *University_President* interacts with the *Science_College* component through the *College_Teach_Algebra* operation call interaction.

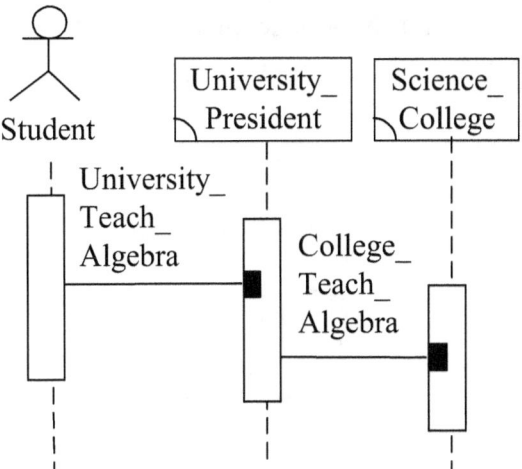

Student

University_
President

Science_
College

University_
Teach_
Algebra

College_
Teach_
Algebra

We draw the infinite-queue SBC process algebra Backus-Naur Form tree of the concept view of *Kurdi University* as follows:

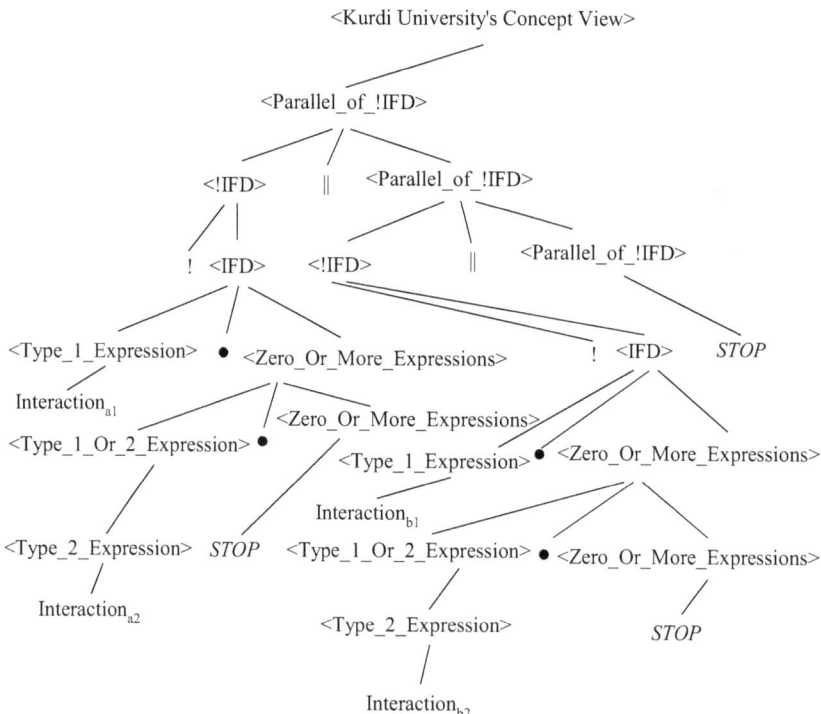

Interaction$_{a1}$ stands for the 1st interaction of the ath interaction flow diagram of the concept view of *Kurdi University*. Interaction$_{a1}$ is a type_1 interaction which describes the *Student* actor interacts with the *University_President* component.

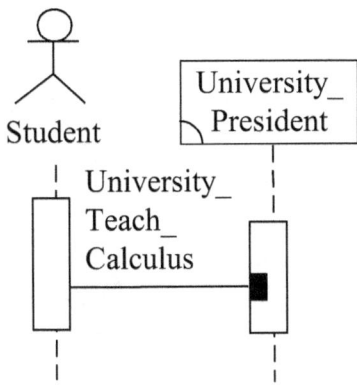

Interaction$_{a2}$ stands for the 2nd interaction of the ath interaction flow diagram of the concept view of *Kurdi University*. Interaction$_{a2}$ is a type_2 interaction which describes the *University_President* component interacts with the *Science_College* component.

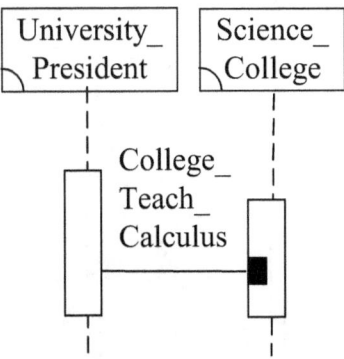

Interaction$_{b1}$ stands for the 1st interaction of the bth interaction flow diagram of the concept view of *Kurdi University*. Interaction$_{b1}$ is a type_1 interaction which describes the *Student* actor interacts with the *University_President* component.

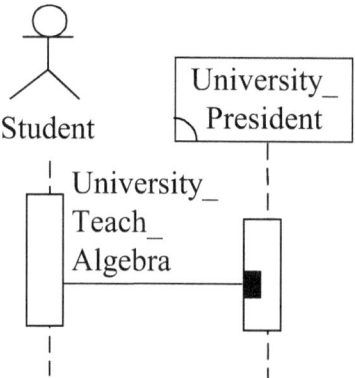

Interaction$_{b2}$ stands for the 2nd interaction of the bth interaction flow diagram of the concept view of *Kurdi University*. Interaction$_{b2}$ is a type_2 interaction which describes the *University_President* component interacts with the *Science_College* component.

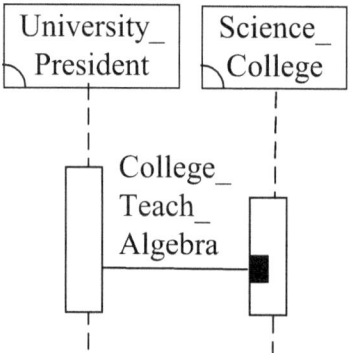

IFD$_a$ describes the ath interaction flow diagram, i.e. *Study_Calculus_Course* behavior, of the concept view of *Kurdi University*. IFD$_a$ is syntactically represented as "Interaction$_{a1}$●Interaction$_{a2}$●*STOP*".

$$IFD_a =$$
$$Interaction_{a1} \bullet Interaction_{a2} \bullet STOP$$

IFD_b describes the bth interaction flow diagram, i.e. *Study_Algebra_Course* behavior, of the concept view of *Kurdi University*. IFD_b is syntactically represented as "$Interaction_{b1} \bullet Interaction_{b2} \bullet STOP$".

$$IFD_b =$$
$$Interaction_{b1} \bullet Interaction_{b2} \bullet STOP$$

Each interaction flow diagram may replicate itself a countably infinite time. $Interaction_{xyz}$ stands for the yth replication of the zth interaction of the xth interaction flow diagram. $!IFD_a$ describes the infinite replication of the ath interaction flow diagram, i.e. *Study_Calculus_Course* behavior, of the concept view of *Kurdi University*. $!IFD_a$ is syntactically represented as "$(Interaction_{a11} \bullet Interaction_{a12} \bullet STOP)$ ||
$(Interaction_{a21} \bullet Interaction_{a22} \bullet STOP)$ ||
$(Interaction_{a31} \bullet Interaction_{a32} \bullet STOP)$ .... ||
$(Interaction_{a\infty1} \bullet Interaction_{a\infty2} \bullet STOP)$".

98

!IFD$_a$ =

(Interaction$_{a11}$•Interaction$_{a12}$•*STOP*)

‖

(Interaction$_{a21}$•Interaction$_{a22}$•*STOP*)

‖

(Interaction$_{a31}$•Interaction$_{a32}$•*STOP*)

....

‖

(Interaction$_{a\infty1}$•Interaction$_{a\infty2}$•*STOP*)

!IFD$_b$ describes the infinite replication of the bth interaction flow diagram, i.e. *Study_Algebra_Course* behavior, of the concept view of *Kurdi University*. !IFD$_b$ is syntactically represented as "(Interaction$_{b11}$•Interaction$_{b12}$•*STOP*)   ‖
(Interaction$_{b21}$•Interaction$_{b22}$•*STOP*)   ‖
(Interaction$_{b31}$•Interaction$_{b32}$•*STOP*)   ....   ‖
(Interaction$_{b\infty1}$•Interaction$_{b\infty2}$•*STOP*)".

!IFD$_b$ =

(Interaction$_{b11}$•Interaction$_{b12}$•*STOP*)

‖

(Interaction$_{b21}$•Interaction$_{b22}$•*STOP*)

‖

(Interaction$_{b31}$•Interaction$_{b32}$•*STOP*)

....

‖

(Interaction$_{b\infty1}$•Interaction$_{b\infty2}$•*STOP*)

Infinite-queue SBC process of the concept view of *Kurdi University* is syntactically represented as $(!IFD_a)\|(!IFD_b)$ which equals to "(Interaction$_{a11}$●Interaction$_{a12}$●*STOP*) $\|$
(Interaction$_{a21}$●Interaction$_{a22}$●*STOP*) $\|$
(Interaction$_{a31}$●Interaction$_{a32}$●*STOP*)   .... $\|$
(Interaction$_{a\infty1}$●Interaction$_{a\infty2}$●*STOP*) $\|$
(Interaction$_{b11}$●Interaction$_{b12}$●*STOP*) $\|$
(Interaction$_{b21}$●Interaction$_{b22}$●*STOP*) $\|$
(Interaction$_{b31}$●Interaction$_{b32}$●*STOP*)   .... $\|$
(Interaction$_{b\infty1}$●Interaction$_{b\infty2}$●*STOP*)".

Kurdi University's Concept View $\overset{\text{def}}{=}$

(Interaction$_{a11}$●Interaction$_{a12}$●*STOP*)
    $\|$
(Interaction$_{a21}$●Interaction$_{a22}$●*STOP*)
    $\|$
(Interaction$_{a31}$●Interaction$_{a32}$●*STOP*) ....
    $\|$
(Interaction$_{a\infty1}$●Interaction$_{a\infty2}$●*STOP*)
$\|$
(Interaction$_{b11}$●Interaction$_{b12}$●*STOP*)
    $\|$
(Interaction$_{b21}$●Interaction$_{b22}$●*STOP*)
    $\|$
(Interaction$_{b31}$●Interaction$_{b32}$●*STOP*) ....
    $\|$
(Interaction$_{b\infty1}$●Interaction$_{b\infty2}$●*STOP*)

Infinite-Queue SBC Process of the Analysis View

We draw the Architecture Hierarchy Diagram (AHD) of the analysis view of *Kurdi University* as follows:

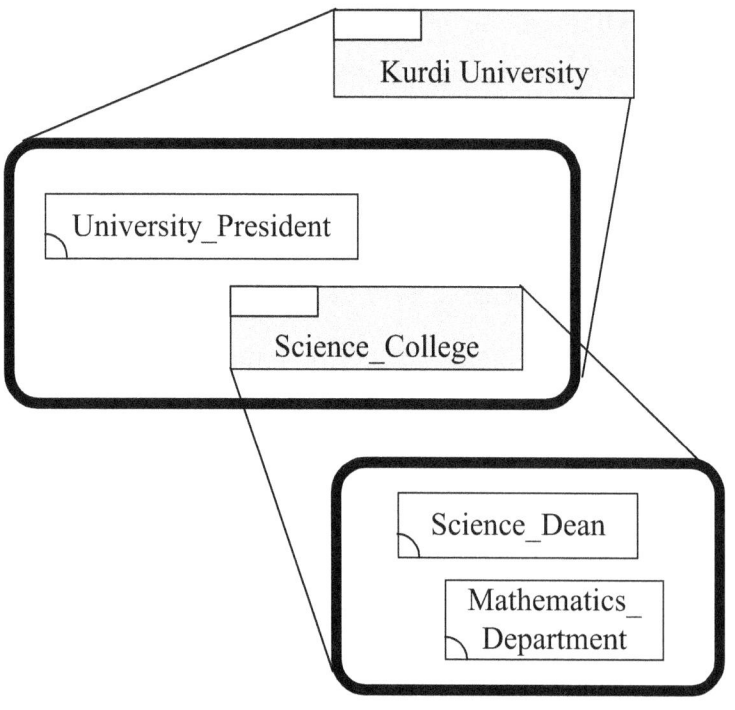

The overall behavior of the analysis view of *Kurdi University* includes two behaviors: *Study_Calculus_Course* and *Study_Algebra_Course*. Each of them is described by an individual IFD.

An IFD of the *Study_Calculus_Course* behavior is shown below. First, actor *Student* interacts with the *University_President* component through the *University_Teach_Calculus* operation call interaction. Next, component *University_President* interacts with the *Science_Dean* component through the *College_Teach_Calculus* operation call interaction. Finally, component *Science_Dean* interacts with the *Mathematics_Department* component through the *Department_Teach_Calculus* operation call interaction.

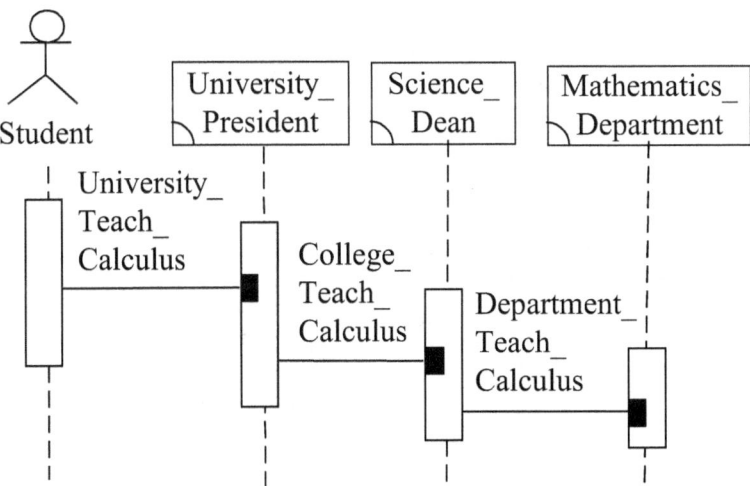

An IFD of the *Study_Algebra_Course* behavior is shown below. First, actor *Student* interacts with the *University_President* component through the *University_Teach_Algebra* operation call interaction. Next, component *University_President* interacts with the *Science_Dean* component through the *College_Teach_Algebra* operation call interaction. Finally, component *Science_Dean* interacts with the *Mathematics_Department* component through the *Department_Teach_Algebra* operation call interaction.

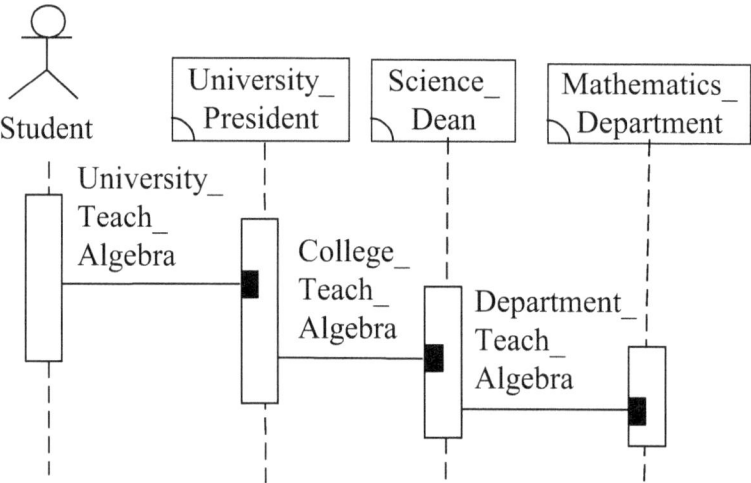

We draw the infinite-queue SBC process algebra Backus-Naur Form tree of the analysis view of *Kurdi University* as follows:

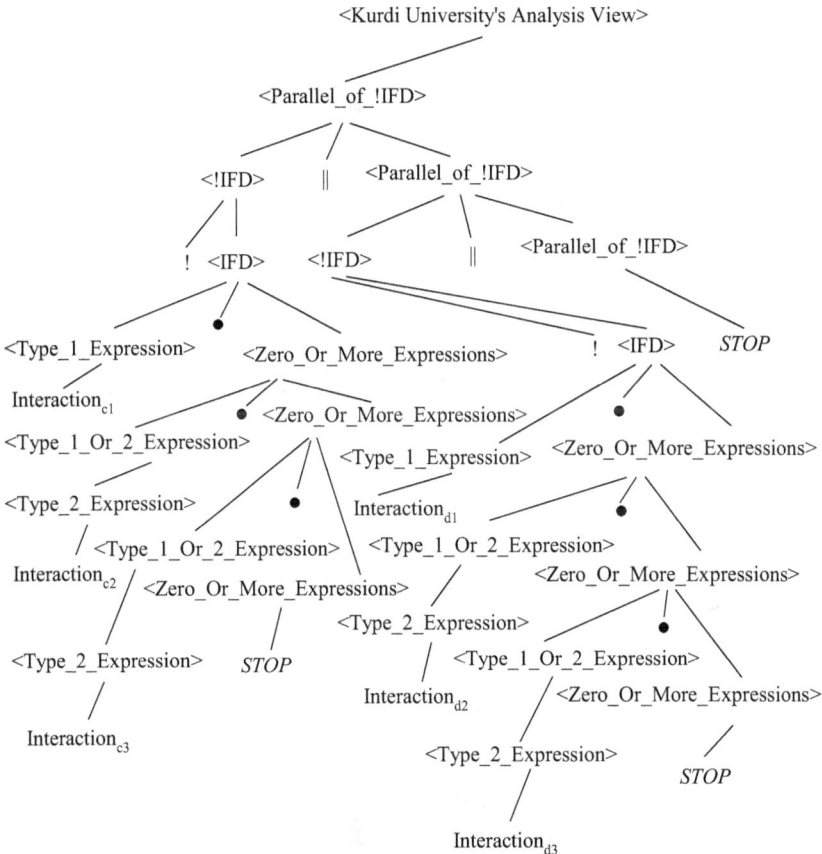

Interaction$_{c1}$ stands for the 1st interaction of the cth interaction flow diagram of the analysis view of *Kurdi University*. Interaction$_{c1}$ is a type_1 interaction which describes the *Student* actor interacts with the *University_President* component.

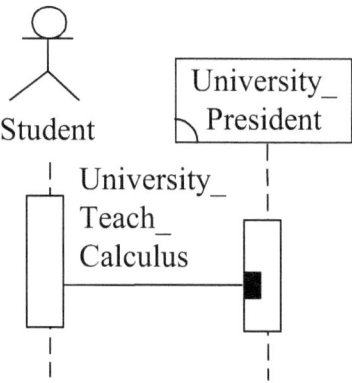

Interaction$_{c2}$ stands for the 2nd interaction of the cth interaction flow diagram of the analysis view of *Kurdi University*. Interaction$_{c2}$ is a type_2 interaction which describes the *University_President* component interacts with the *Science_Dean* component.

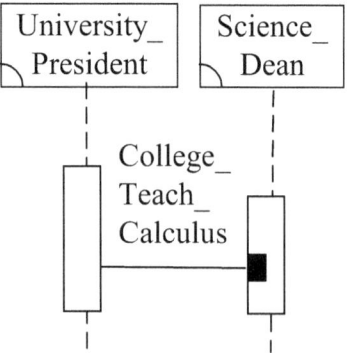

Interaction$_{c3}$ stands for the 3rd interaction of the cth interaction flow diagram of the analysis view of *Kurdi University*. Interaction$_{c3}$ is a type_2 interaction which describes the *Science_Dean* component interacts with the *Mathematics_Department* component.

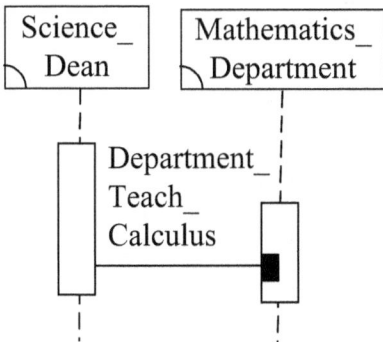

Interaction$_{d1}$ stands for the 1st interaction of the dth interaction flow diagram of the analysis view of *Kurdi University*. Interaction$_{d1}$ is a type_1 interaction which describes the *Student* actor interacts with the *University_President* component.

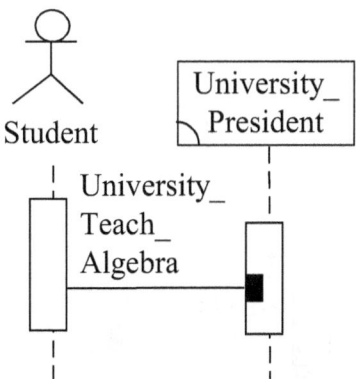

Interaction$_{d2}$ stands for the 2nd interaction of the dth interaction flow diagram of the analysis view of *Kurdi University*. Interaction$_{d2}$ is a type_2 interaction which describes the *University_President* component interacts with the *Science_Dean* component.

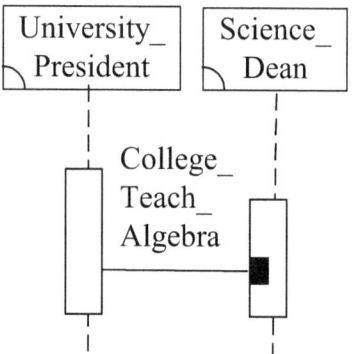

Interaction$_{d3}$ stands for the 3rd interaction of the dth interaction flow diagram of the analysis view of *Kurdi University*. Interaction$_{d3}$ is a type_2 interaction which describes the *Science_Dean* component interacts with the *Mathematics_Department* component.

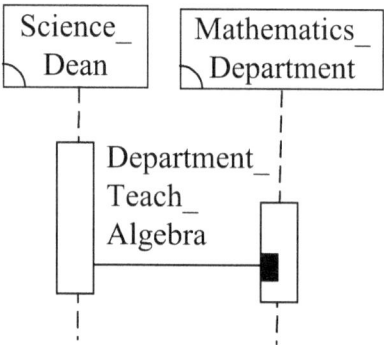

IFD$_c$ describes the cth interaction flow diagram, i.e. *Study_Calculus_Course* behavior, of the analysis view of *Kurdi University*. IFD$_c$ is syntactically represented as "Interaction$_{c1}$•Interaction$_{c2}$•Interaction$_{c3}$•*STOP*".

$$IFD_c =$$

$$Interaction_{c1} \bullet Interaction_{c2} \bullet Interaction_{c3} \bullet STOP$$

IFD_d describes the dth interaction flow diagram, i.e. *Study_Algebra_Course* behavior, of the analysis view of *Kurdi University*. IFD_d is syntactically represented as "$Interaction_{d1} \bullet Interaction_{d2} \bullet Interaction_{d3} \bullet STOP$".

$$IFD_d =$$

$$Interaction_{d1} \bullet Interaction_{d2} \bullet Interaction_{d3} \bullet STOP$$

Each interaction flow diagram may replicate itself a countably infinite time. $Interaction_{xyz}$ stands for the yth replication of the zth interaction of the xth interaction flow diagram. $!IFD_c$ describes the infinite replication of the cth interaction flow diagram, i.e. *Study_Calculus_Course* behavior, of the analysis view of *Kurdi University*. $!IFD_c$ is syntactically represented as "$(Interaction_{c11} \bullet Interaction_{c12} \bullet Interaction_{c13} \bullet STOP)$ $\parallel$ $(Interaction_{c21} \bullet Interaction_{c22} \bullet Interaction_{c23} \bullet STOP)$ $\parallel$ $(Interaction_{c31} \bullet Interaction_{c32} \bullet Interaction_{c33} \bullet STOP)$ .... $\parallel$ $(Interaction_{c\infty1} \bullet Interaction_{c\infty2} \bullet Interaction_{c\infty3} \bullet STOP)$".

$!IFD_c =$

$(Interaction_{c11} \bullet Interaction_{c12} \bullet Interaction_{c13} \bullet STOP)$

$\parallel$

$(Interaction_{c21} \bullet Interaction_{c22} \bullet Interaction_{c23} \bullet STOP)$

$\parallel$

$(Interaction_{c31} \bullet Interaction_{c32} \bullet Interaction_{c33} \bullet STOP)$

....

$\parallel$

$(Interaction_{c\infty1} \bullet Interaction_{c\infty2} \bullet Interaction_{c\infty3} \bullet STOP)$

$!IFD_d$ describes the infinite replication of the dth interaction flow diagram, i.e. *Study_Algebra_Course* behavior, of the analysis view of *Kurdi University*. $!IFD_d$ is syntactically represented as "$(Interaction_{d11} \bullet Interaction_{d12} \bullet Interaction_{d13} \bullet STOP)$ $\parallel$ $(Interaction_{d21} \bullet Interaction_{d22} \bullet Interaction_{d23} \bullet STOP)$ $\parallel$ $(Interaction_{d31} \bullet Interaction_{d32} \bullet Interaction_{d33} \bullet STOP)$ .... $\parallel$ $(Interaction_{d\infty1} \bullet Interaction_{d\infty2} \bullet Interaction_{d\infty3} \bullet STOP)$".

$!IFD_d =$

$(Interaction_{d11} \bullet Interaction_{d12} \bullet Interaction_{d13} \bullet STOP)$

$\parallel$

$(Interaction_{d21} \bullet Interaction_{d22} \bullet Interaction_{d23} \bullet STOP)$

$\parallel$

$(Interaction_{d31} \bullet Interaction_{d32} \bullet Interaction_{d33} \bullet STOP)$

....

$\parallel$

$(Interaction_{d\infty1} \bullet Interaction_{d\infty2} \bullet Interaction_{d\infty3} \bullet STOP)$

Infinite-queue SBC process of the analysis view of *Kurdi University* is syntactically represented as $(!IFD_c)\|(!IFD_d)$ which equals to "(Interaction$_{c11}$●Interaction$_{c12}$●Interaction$_{c13}$●*STOP*) ‖ (Interaction$_{c21}$●Interaction$_{c22}$●Interaction$_{c23}$●*STOP*) ‖ (Interaction$_{c31}$●Interaction$_{c32}$●Interaction$_{c33}$●*STOP*) .... ‖ (Interaction$_{c\infty1}$●Interaction$_{c\infty2}$●Interaction$_{c\infty3}$●*STOP*) ‖ (Interaction$_{d11}$●Interaction$_{d12}$●Interaction$_{d13}$●*STOP*) ‖ (Interaction$_{d21}$●Interaction$_{d22}$●Interaction$_{d23}$●*STOP*) ‖ (Interaction$_{d31}$●Interaction$_{d32}$●Interaction$_{d33}$●*STOP*) .... ‖ (Interaction$_{d\infty1}$●Interaction$_{d\infty2}$●Interaction$_{d\infty3}$●*STOP*)".

Kurdi University's Analysis View $\overset{def}{=\!=}$

(Interaction$_{c11}$●Interaction$_{c12}$●Interaction$_{c13}$●*STOP*)
‖
(Interaction$_{c21}$●Interaction$_{c22}$●Interaction$_{c23}$●*STOP*)
‖
(Interaction$_{c31}$●Interaction$_{c32}$●Interaction$_{c33}$●*STOP*) ....
‖
(Interaction$_{c\infty1}$●Interaction$_{c\infty2}$●Interaction$_{c\infty3}$●*STOP*)
‖
(Interaction$_{d11}$●Interaction$_{d12}$●Interaction$_{d13}$●*STOP*)
‖
(Interaction$_{d21}$●Interaction$_{d22}$●Interaction$_{d23}$●*STOP*)
‖
(Interaction$_{d31}$●Interaction$_{d32}$●Interaction$_{d33}$●*STOP*) ....
‖
(Interaction$_{d\infty1}$●Interaction$_{d\infty2}$●Interaction$_{d\infty3}$●*STOP*)

Infinite-Queue SBC Process of the Structural Composition of the Analysis View

Structural composition of the analysis view of *Kurdi University* means to compose the *Science_Dean* and *Mathematics_Department* components into the *Science_College* component. That is, we will rename the *Science_Dean* component to the *Science_College* component; we also will rename the *Mathematics_Department* component to the *Science_College* component.

[Science_College/Science_Dean,
Science_College/Mathematics_Department]

We draw the infinite-queue SBC process algebra Backus-Naur Form tree of the structural composition of the analysis view of *Kurdi University* as follows:

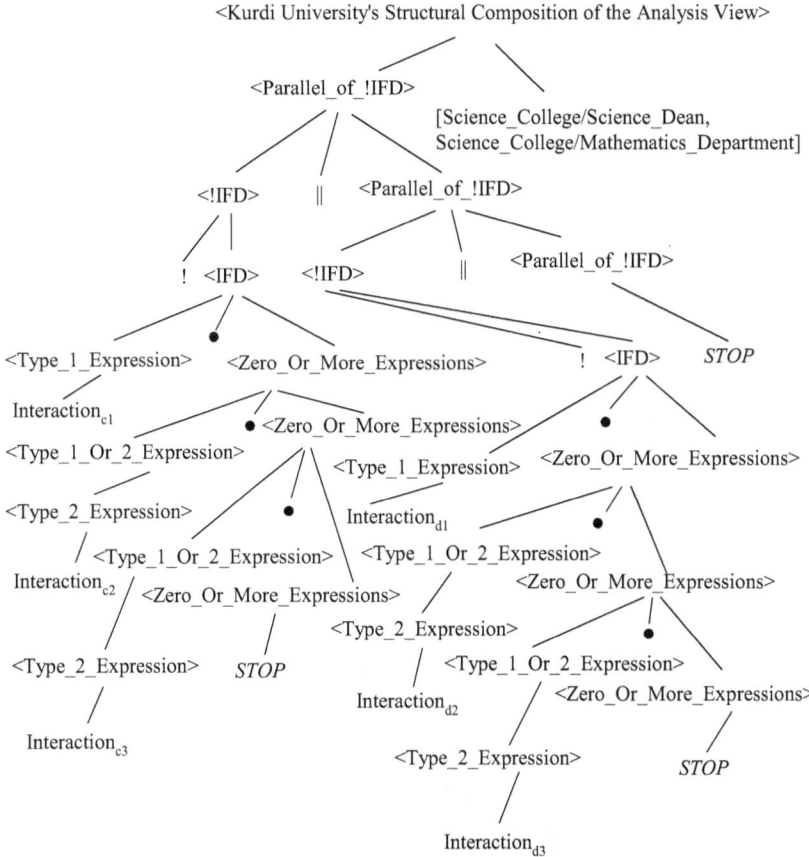

Interaction$_{e1}$=Interaction$_{c1}$[Science_College/Science_Dean,Science_College/Mathematics_Department]=Interaction$_{a1}$ stands for the 1st interaction of the eth interaction flow diagram of the structural composition of the analysis view of *Kurdi University*. Interaction$_{e1}$ is a type_1 interaction which describes the *Student* actor interacts with the *University_President* component.

112

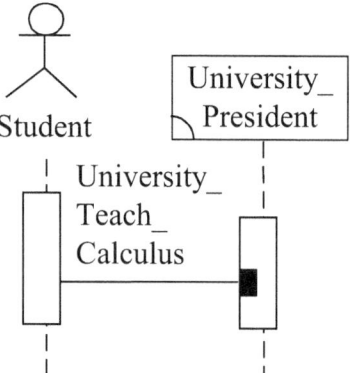

Interaction$_{e2}$=Interaction$_{c2}$[Science_College/Science_Dean,S cience_College/Mathematics_Department]=Interaction$_{a2}$ stands for the 2nd interaction of the eth interaction flow diagram of the structural composition of the analysis view of *Kurdi University*. Interaction$_{e2}$ is a type_2 interaction which describes the *University_President* component interacts with the *Science_College* component.

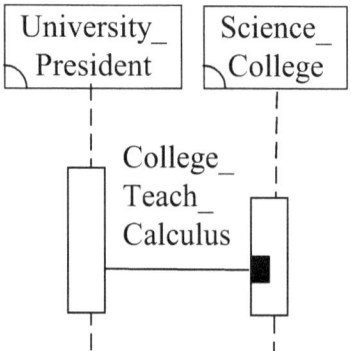

Interaction$_{e3}$=Interaction$_{c3}$[Science_College/Science_Dean,S cience_College/Mathematics_Department] stands for the 3rd interaction of the eth interaction flow diagram of the structural composition of the analysis view of *Kurdi University*. Interaction$_{e3}$ is an internal interaction (i.e. λ) inside the *Science_College* component.

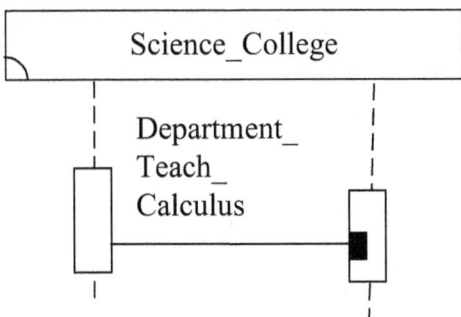

Interaction$_{f1}$=Interaction$_{d1}$[Science_College/Science_Dean,Science_College/Mathematics_Department]=Interaction$_{b1}$ stands for the 1st interaction of the fth interaction flow diagram of the structural composition of the analysis view of *Kurdi University*. Interaction$_{f1}$ is a type_1 interaction which describes the *Student* actor interacts with the *University_President* component.

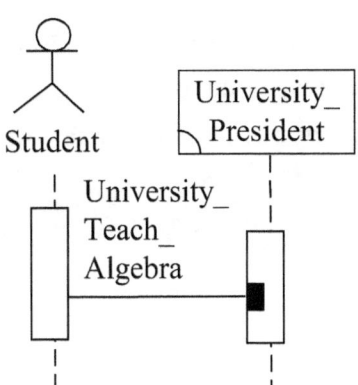

Interaction$_{f2}$=Interaction$_{d2}$[Science_College/Science_Dean,S cience_College/Mathematics_Department]=Interaction$_{b2}$ stands for the 2nd interaction of the fth interaction flow diagram of the structural composition of the analysis view of *Kurdi University*. Interaction$_{f2}$ is a type_2 interaction which describes the *University_President* component interacts with the *Science_College* component.

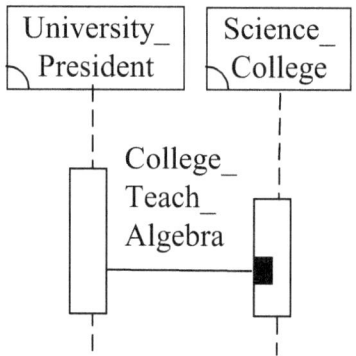

Interaction$_{f3}$=Interaction$_{d3}$[Science_College/Science_Dean,S cience_College/Mathematics_Department] stands for the 3rd interaction of the fth interaction flow diagram of the structural composition of the analysis view of *Kurdi University*. Interaction$_{f3}$ is an internal interaction (i.e. λ) inside the *Science_College* component.

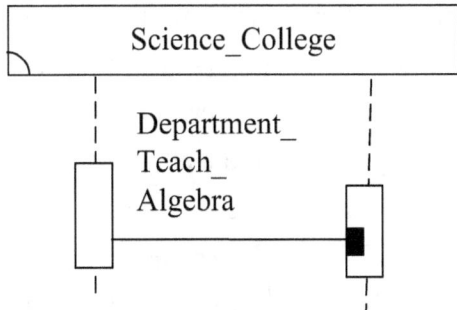

IFD$_e$ describes the eth interaction flow diagram, i.e. *Study_Calculus_Course* behavior, of the structural composition of the analysis view of *Kurdi University*. IFD$_e$ is syntactically represented as "Interaction$_{a1}$●Interaction$_{a2}$●λ●*STOP*".

IFD$_e$ =

Interaction$_{a1}$●Interaction$_{a2}$●λ●*STOP*

IFD$_f$ describes the fth interaction flow diagram, i.e. *Study_Algebra_Course* behavior, of the structural composition of the analysis view of *Kurdi University*. IFD$_f$ is syntactically represented as "Interaction$_{b1}$●Interaction$_{b2}$●λ●*STOP*".

IFD$_f$ =

Interaction$_{b1}$●Interaction$_{b2}$●λ●*STOP*

Each interaction flow diagram may replicate itself a countably infinite time. Interaction$_{xyz}$ stands for the yth replication of the zth interaction of the xth interaction flow diagram. !IFD$_e$ describes the infinite replication of the eth interaction flow diagram, i.e. *Study_Calculus_Course* behavior, of the structural composition of the analysis view of *Kurdi University*. !IFD$_e$ is syntactically represented as "(Interaction$_{a11}$•Interaction$_{a12}$•λ•*STOP*) || (Interaction$_{a21}$•Interaction$_{a22}$•λ•*STOP*) || (Interaction$_{a31}$•Interaction$_{a32}$•λ•*STOP*) .... || (Interaction$_{a\infty1}$•Interaction$_{a\infty2}$•λ•*STOP*)".

$$
\begin{aligned}
&!IFD_e = \\[4pt]
&(Interaction_{a11} \bullet Interaction_{a12} \bullet \lambda \bullet STOP) \\
&\quad || \\
&(Interaction_{a21} \bullet Interaction_{a22} \bullet \lambda \bullet STOP) \\
&\quad || \\
&(Interaction_{a31} \bullet Interaction_{a32} \bullet \lambda \bullet STOP) \\
&\qquad .... \\
&\quad || \\
&(Interaction_{a\infty1} \bullet Interaction_{a\infty2} \bullet \lambda \bullet STOP)
\end{aligned}
$$

!IFD$_f$ describes the infinite replication of the eth interaction flow diagram, i.e. *Study_Algebra_Course* behavior, of the structural composition of the analysis view of *Kurdi University*. !IFD$_f$ is syntactically represented as "(Interaction$_{b11}$•Interaction$_{b12}$•λ•*STOP*) || (Interaction$_{b21}$•Interaction$_{b22}$•λ•*STOP*) || (Interaction$_{b31}$•Interaction$_{b32}$•λ•*STOP*) .... ||

$(\text{Interaction}_{b\infty1} \bullet \text{Interaction}_{b\infty2} \bullet \lambda \bullet STOP)$".

$!\text{IFD}_f =$

$(\text{Interaction}_{b11} \bullet \text{Interaction}_{b12} \bullet \lambda \bullet STOP)$
$\parallel$
$(\text{Interaction}_{b21} \bullet \text{Interaction}_{b22} \bullet \lambda \bullet STOP)$
$\parallel$
$(\text{Interaction}_{b31} \bullet \text{Interaction}_{b32} \bullet \lambda \bullet STOP)$
....
$\parallel$
$(\text{Interaction}_{b\infty1} \bullet \text{Interaction}_{b\infty2} \bullet \lambda \bullet STOP)$

Infinite-queue SBC process of the structural composition of the analysis view of *Kurdi University* is syntactically represented as $(!\text{IFD}_e)\|(!\text{IFD}_f)$ which equals to
"$(\text{Interaction}_{a11} \bullet \text{Interaction}_{a12} \bullet \lambda \bullet STOP)$ $\parallel$
$(\text{Interaction}_{a21} \bullet \text{Interaction}_{a22} \bullet \lambda \bullet STOP)$ $\parallel$
$(\text{Interaction}_{a31} \bullet \text{Interaction}_{a32} \bullet \lambda \bullet STOP)$ .... $\parallel$
$(\text{Interaction}_{a\infty1} \bullet \text{Interaction}_{a\infty2} \bullet \lambda \bullet STOP)$ $\parallel$
$(\text{Interaction}_{b11} \bullet \text{Interaction}_{b12} \bullet \lambda \bullet STOP)$ $\parallel$
$(\text{Interaction}_{b21} \bullet \text{Interaction}_{b22} \bullet \lambda \bullet STOP)$ $\parallel$
$(\text{Interaction}_{b31} \bullet \text{Interaction}_{b32} \bullet \lambda \bullet STOP)$ .... $\parallel$
$(\text{Interaction}_{b\infty1} \bullet \text{Interaction}_{b\infty2} \bullet \lambda \bullet STOP)$".

6

Kurdi University's Structural Composition of the Analysis View $\overset{\text{def}}{=\!=}$

$(\text{Interaction}_{a11} \bullet \text{Interaction}_{a12} \bullet \lambda \bullet STOP)$

$\parallel$

$(\text{Interaction}_{a21} \bullet \text{Interaction}_{a22} \bullet \lambda \bullet STOP)$

$\parallel$

$(\text{Interaction}_{a31} \bullet \text{Interaction}_{a32} \bullet \lambda \bullet STOP)$

....

$\parallel$

$(\text{Interaction}_{a\infty1} \bullet \text{Interaction}_{a\infty2} \bullet \lambda \bullet STOP)$

$\parallel$

$(\text{Interaction}_{b11} \bullet \text{Interaction}_{b12} \bullet \lambda \bullet STOP)$

$\parallel$

$(\text{Interaction}_{b21} \bullet \text{Interaction}_{b22} \bullet \lambda \bullet STOP)$

$\parallel$

$(\text{Interaction}_{b31} \bullet \text{Interaction}_{b32} \bullet \lambda \bullet STOP)$

....

$\parallel$

$(\text{Interaction}_{b\infty1} \bullet \text{Interaction}_{b\infty2} \bullet \lambda \bullet STOP)$

Observation Congruence of "the Concept View" and "the Structural Composition of the Analysis View"

We syntactically represent infinite-queue SBC process P_{01} as $IFD_a \| IFD_b$ which equals to (Interaction$_{a1}$●Interaction$_{a2}$●$STOP$)$\|$ (Interaction$_{b1}$●Interaction$_{b2}$●$STOP$). We also syntactically represent infinite-queue SBC processes P_{02} as (Interaction$_{a2}$●$STOP$)$\|$(Interaction$_{b1}$●Interaction$_{b2}$●$STOP$), P_{03} as (Interaction$_{b1}$●Interaction$_{b2}$●$STOP$), P_{04} as (Interaction$_{a1}$●Interaction$_{a2}$●$STOP$)$\|$(Interaction$_{b2}$●$STOP$), P_{05} as (Interaction$_{a2}$●$STOP$)$\|$(Interaction$_{b2}$●$STOP$), P_{06} as (Interaction$_{b2}$●$STOP$), P_{07} as (Interaction$_{a1}$●Interaction$_{a2}$●$STOP$), P_{08} as (Interaction$_{a2}$●$STOP$), P_{09} as $STOP$, respectively.

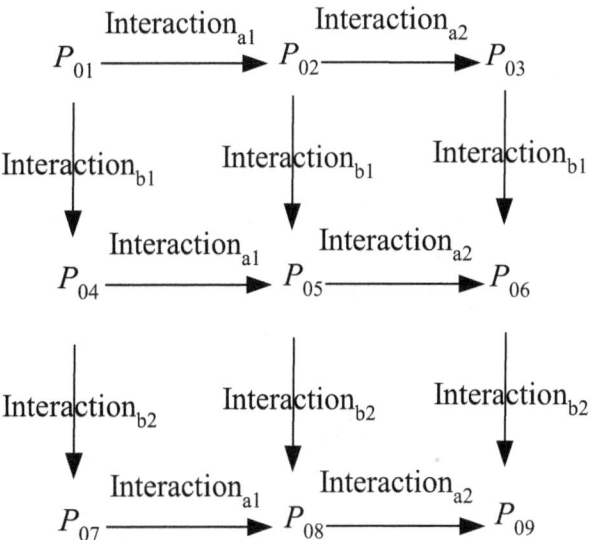

We syntactically represent infinite-queue SBC process Q_{01} as $IFD_e \| IFD_f$ which equals to $(Interaction_{a1} \bullet Interaction_{a2} \bullet \lambda \bullet STOP) \| (Interaction_{b1} \bullet Interaction_{b2} \bullet \lambda \bullet STOP)$. We also syntactically represent infinite-queue SBC processes Q_{02} as $(Interaction_{a2} \bullet \lambda \bullet STOP) \| (Interaction_{b1} \bullet Interaction_{b2} \bullet \lambda \bullet STOP)$, Q_{03} as $(\lambda \bullet STOP) \| (Interaction_{b1} \bullet Interaction_{b2} \bullet \lambda \bullet STOP)$, Q_{04} as $(Interaction_{b1} \bullet Interaction_{b2} \bullet \lambda \bullet STOP)$, Q_{05} as $(Interaction_{a1} \bullet Interaction_{a2} \bullet \lambda \bullet STOP) \| (Interaction_{b2} \bullet \lambda \bullet STOP)$, Q_{06} as $(Interaction_{a2} \bullet \lambda \bullet STOP) \| (Interaction_{b2} \bullet \lambda \bullet STOP)$, Q_{07} as $(\lambda \bullet STOP) \| (Interaction_{b2} \bullet \lambda \bullet STOP)$, Q_{08} as $(Interaction_{b2} \bullet \lambda \bullet STOP)$, Q_{09} as $(Interaction_{a1} \bullet Interaction_{a2} \bullet \lambda \bullet STOP) \| (\lambda \bullet STOP)$, Q_{10} as $(Interaction_{a2} \bullet \lambda \bullet STOP) \| (\lambda \bullet STOP)$, Q_{11} as $(\lambda \bullet STOP) \| (\lambda \bullet STOP)$, Q_{12} as $(\lambda \bullet STOP)$, Q_{13} as $(Interaction_{a1} \bullet Interaction_{a2} \bullet \lambda \bullet STOP)$, Q_{14} as $(Interaction_{a2} \bullet \lambda \bullet STOP)$, Q_{15} as $(\lambda \bullet STOP)$, Q_{16} as $STOP$, respectively.

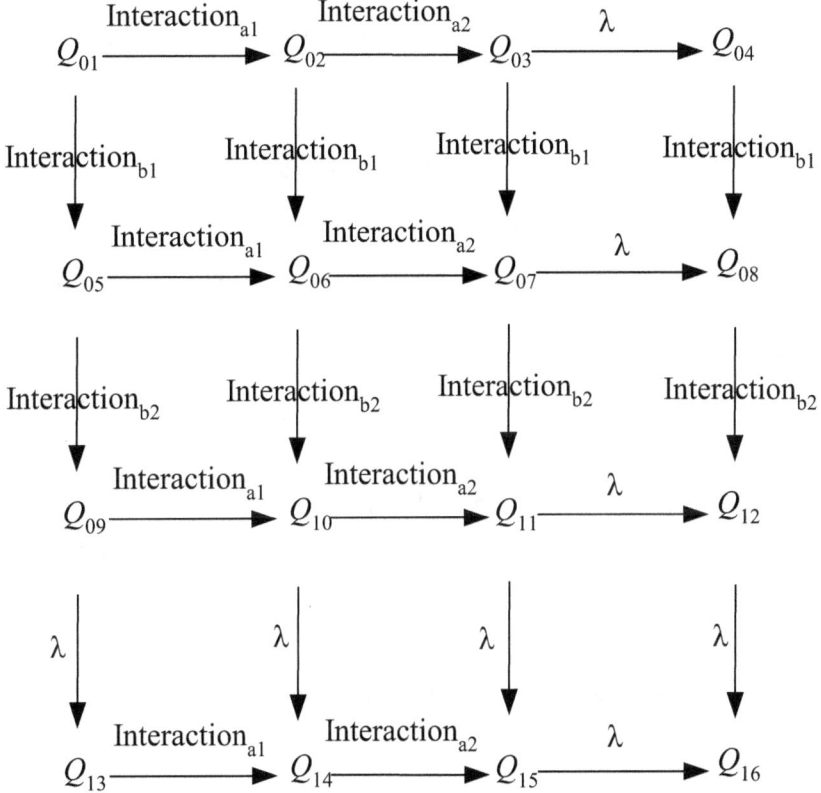

We can easily verify that $S = \{(P_{01}, Q_{01}), (P_{02}, Q_{02}), (P_{03}, Q_{03}), (P_{03}, Q_{04}), (P_{04}, Q_{05}), (P_{05}, Q_{06}), (P_{06}, Q_{07}), (P_{06}, Q_{08}), (P_{07}, Q_{09}), (P_{07}, Q_{13}), (P_{08}, Q_{10}), (P_{08}, Q_{14}), (P_{09}, Q_{11}), (P_{09}, Q_{12}), (P_{09}, Q_{15}), (P_{09}, Q_{16})\}$ is a bisimulation.

Using the S bisimulation, we then are able to verify that P_{01} and Q_{01} are observation congruent because (1)

$$P_{01} \xrightarrow{\text{Interaction}_{a1}} P_{02}, \text{ then we have } Q_{02} \text{ that } Q_{01} \overset{\text{Interaction}_{a1}}{\Longrightarrow} Q_{02}$$

and $P_{02} \overset{\sim}{\approx} Q_{02}$, and (2) $P_{01} \xrightarrow{\text{Interaction}_{b1}} P_{04}$, then we have Q_{05} that

$$Q_{01} \overset{\text{Interaction}_{b1}}{\Longrightarrow} Q_{05} \text{ and } P_{04} \overset{\sim}{\approx} Q_{05}, \text{ and (3) } Q_{01} \xrightarrow{\text{Interaction}_{a1}} Q_{02},$$

then we have P_{02} that $P_{01} \overset{\text{Interaction}_{a1}}{\Longrightarrow} P_{02}$ and $P_{02} \overset{\sim}{\approx} Q_{02}$, and (4)

$$Q_{01} \xrightarrow{\text{Interaction}_{b1}} Q_{05}, \text{ then we have } P_{04} \text{ that } P_{01} \overset{\text{Interaction}_{b1}}{\Longrightarrow} P_{04}$$

and $P_{04} \overset{\sim}{\approx} Q_{05}$.

$P_{01} = Q_{01}$ means $IFD_a \| IFD_b = IFD_e \| IFD_f$, and $IFD_a \| IFD_b = IFD_e \| IFD_f$ means $!(IFD_a \| IFD_b) = !(IFD_e \| IFD_f)$, and $!(IFD_a \| IFD_b) = !(IFD_e \| IFD_f)$ means $(!IFD_a) \| (!IFD_b) = (!IFD_e) \| (!IFD_f)$. So, there is observation congruence of "the concept view of *Kurdi University*" and "the structural composition of the analysis view of *Kurdi University*".

Conclusively, the analysis view of *Kurdi University* is one level down structural decomposition (with observation congruence verification) of the concept view of *Kurdi University*.

Infinite-Queue SBC Process of the Design View

We draw the Architecture Hierarchy Diagram of the design view of *Kurdi University* as follows:

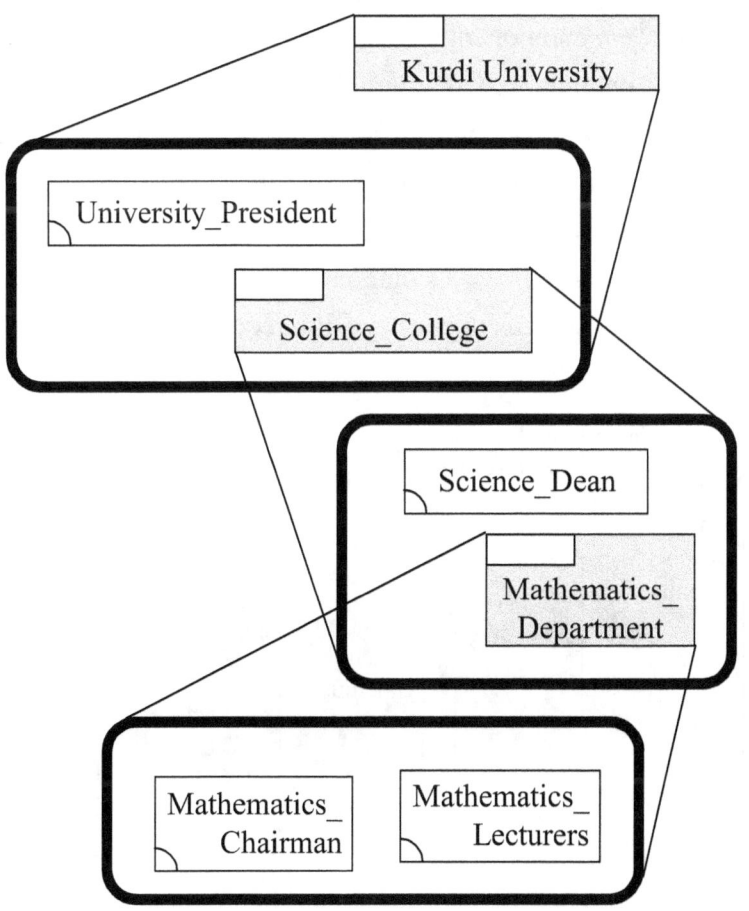

The overall behavior of the design view of *Kurdi University* includes two behaviors: *Study_Calculus_Course* and *Study_Algebra_Course*. Each of them is described by an individual IFD.

An IFD of the *Study_Calculus_Course* behavior is shown below. First, actor *Student* interacts with the *University_President* component through the *University_Teach_Calculus* operation call interaction. Next, component *University_President* interacts with the *Science_Dean* component through the *College_Teach_Calculus* operation call interaction. Continuingly, component *Science_Dean* interacts with the *Mathematics_Chairman* component through the *Department_Teach_Calculus* operation call interaction. Finally, component *Mathematics_Chairman* interacts with the *Mathematics_Lecturers* component through the *Lecturer_Teach_Calculus* operation call interaction.

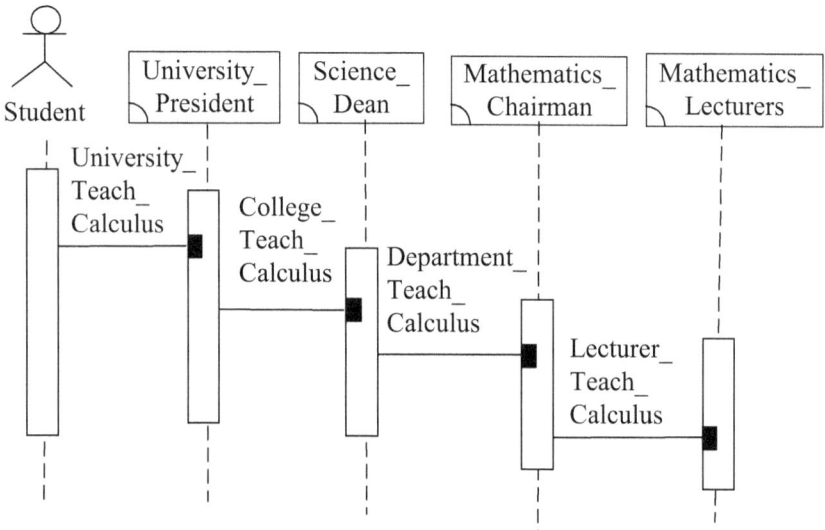

An IFD of the *Study_Algebra_Course* behavior is shown below. First, actor *Student* interacts with the *University_President* component through the *University_Teach_Algebra* operation call interaction. Next, component *University_President* interacts with the *Science_Dean* component through the *College_Teach_Algebra* operation call interaction. Continuingly, component *Science_Dean* interacts with the *Mathematics_Chairman* component through the *Department_Teach_Algebra* operation call interaction. Finally, component *Mathematics_Chairman* interacts with the *Mathematics_Lecturers* component through the *Lecturer_Teach_Algebra* operation call interaction.

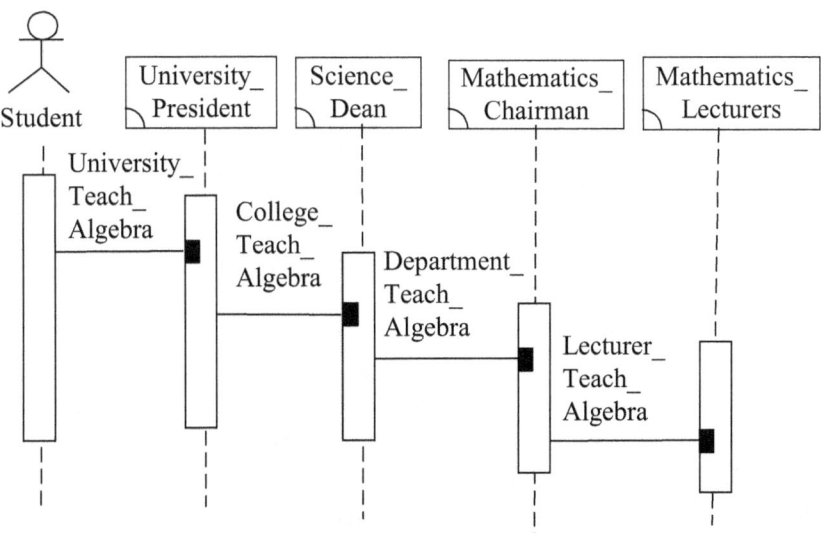

We draw the infinite-queue SBC process algebra Backus-Naur Form tree of the design view of *Kurdi University* as follows:

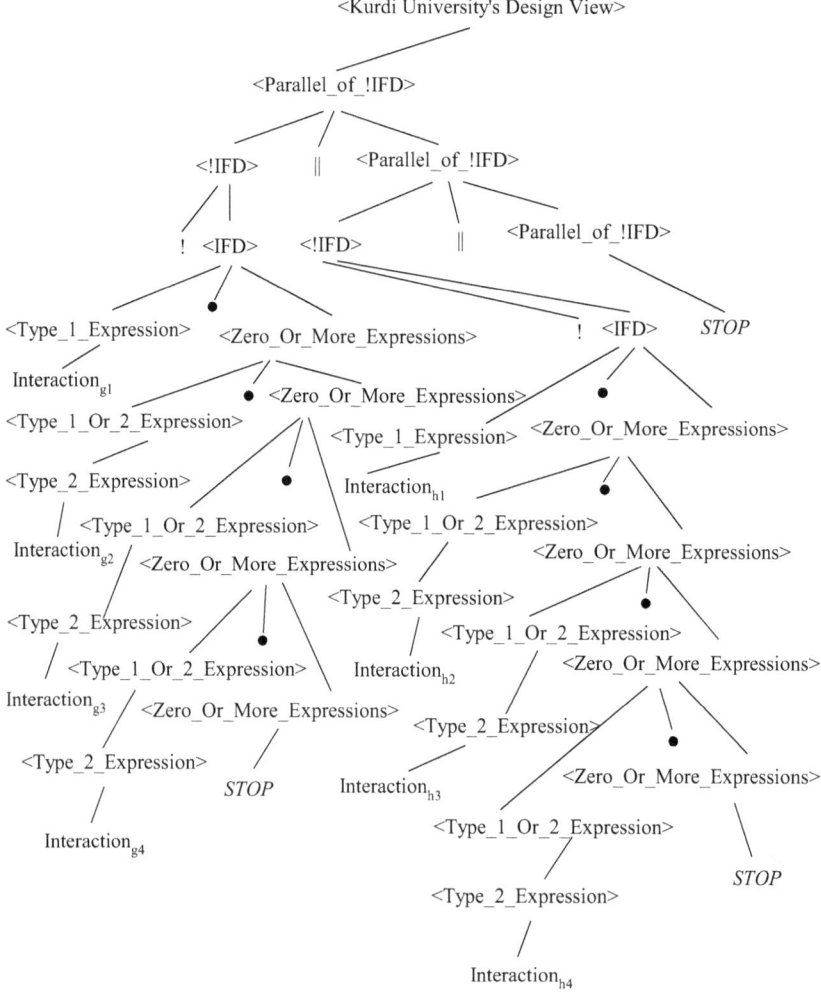

Interaction$_{g1}$ stands for the 1st interaction of the gth interaction flow diagram of the design view of *Kurdi University*. Interaction$_{g1}$ is a type_1 interaction which describes the *Student* actor interacts with the *University_President* component.

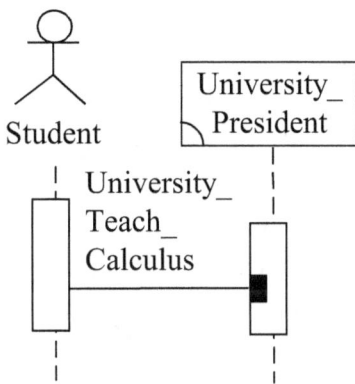

Interaction$_{g2}$ stands for the 2nd interaction of the gth interaction flow diagram of the design view of *Kurdi University*. Interaction$_{g2}$ is a type_2 interaction which describes the *University_President* component interacts with the *Science_Dean* component.

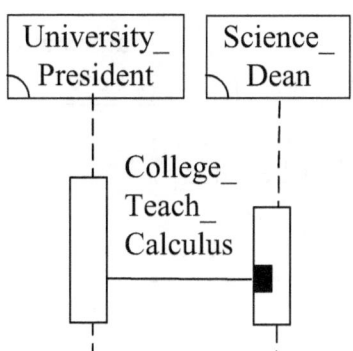

Interaction$_{g3}$ stands for the 3rd interaction of the gth interaction flow diagram of the design view of *Kurdi University*. Interaction$_{g3}$ is a type_2 interaction which describes the *Science_Dean* component interacts with the *Mathematics_Chairman* component.

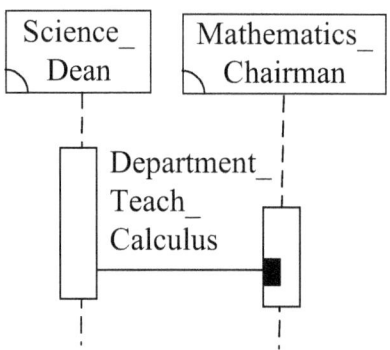

Interaction$_{g4}$ stands for the 4th interaction of the gth interaction flow diagram of the design view of *Kurdi University*. Interaction$_{g4}$ is a type_2 interaction which describes the *Mathematics_Chairman* component interacts with the *Mathematics_Lecturers* component.

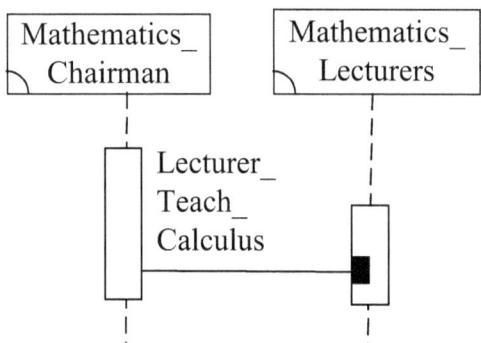

Interaction$_{h1}$ stands for the 1st interaction of the hth interaction flow diagram of the design view of *Kurdi University*. Interaction$_{h1}$ is a type_1 interaction which describes the *Student* actor interacts with the *University_President* component.

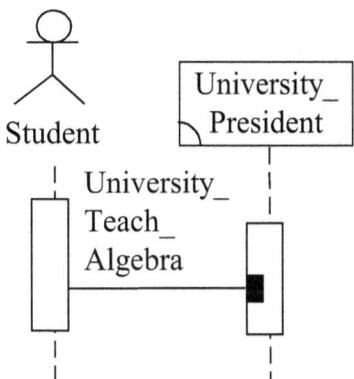

Interaction$_{h2}$ stands for the 2nd interaction of the hth interaction flow diagram of the design view of *Kurdi University*. Interaction$_{h2}$ is a type_2 interaction which describes the *University_President* component interacts with the *Science_Dean* component.

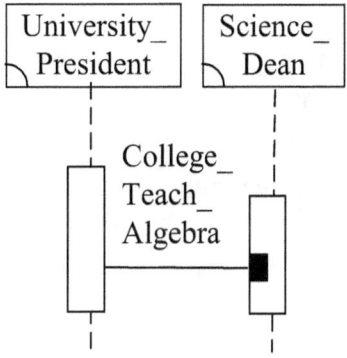

Interaction$_{h3}$ stands for the 3rd interaction of the hth interaction flow diagram of the design view of *Kurdi University*. Interaction$_{h3}$ is a type_2 interaction which describes the *Science_Dean* component interacts with the *Mathematics_Chairman* component.

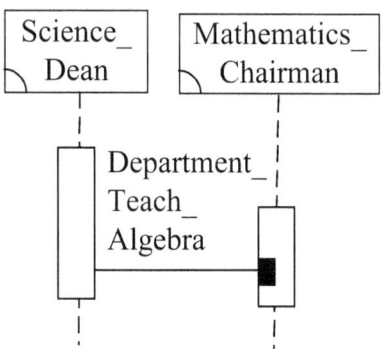

Interaction$_{h4}$ stands for the 4th interaction of the hth interaction flow diagram of the design view of *Kurdi University*. Interaction$_{h4}$ is a type_2 interaction which describes the *Mathematics_Chairman* component interacts with the *Mathematics_Lecturers* component.

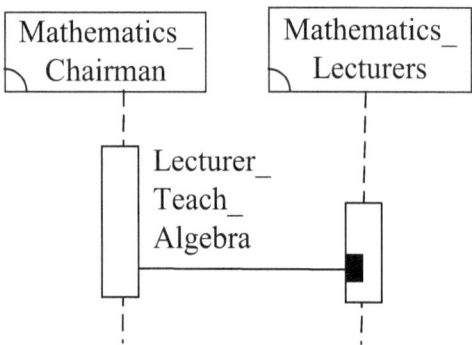

IFD$_g$ describes the gth interaction flow diagram, i.e. *Study_Calculus_Course* behavior, of the design view of *Kurdi University*. IFD$_g$ is syntactically represented as "Interaction$_{g1}$●Interaction$_{g2}$●Interaction$_{g3}$●Interaction$_{g4}$●*STOP*".

$$\text{IFD}_g =$$

$$\text{Interaction}_{g1} \bullet \text{Interaction}_{g2} \bullet \text{Interaction}_{g3} \bullet \text{Interaction}_{g4} \bullet STOP$$

IFD$_h$ describes the hth interaction flow diagram, i.e. *Study_Algebra_Course* behavior, of the design view of *Kurdi University*. IFD$_h$ is syntactically represented as "Interaction$_{h1}$●Interaction$_{h2}$●Interaction$_{h3}$●Interaction$_{h4}$●*STOP*".

$$\text{IFD}_h =$$

$$\text{Interaction}_{h1} \bullet \text{Interaction}_{h2} \bullet \text{Interaction}_{h3} \bullet \text{Interaction}_{h4} \bullet STOP$$

Each interaction flow diagram may replicate itself a countably infinite time. Interaction$_{xyz}$ stands for the yth replication of the zth interaction of the xth interaction flow diagram. !IFD$_g$ describes the infinite replication of the gth interaction flow diagram, i.e. *Study_Calculus_Course* behavior, of the design view of *Kurdi University*. !IFD$_g$ is syntactically represented as "(Interaction$_{g11}$●Interaction$_{g12}$●Interaction$_{g13}$●Interaction$_{g14}$●*STOP*)| |(Interaction$_{g21}$●Interaction$_{g22}$●Interaction$_{g23}$●Interaction$_{g24}$●*STOP*) ||(Interaction$_{g31}$●Interaction$_{g32}$●Interaction$_{g33}$●Interaction$_{g34}$●*STOP*)… ….||(Interaction$_{g\infty1}$●Interaction$_{g\infty2}$●Interaction$_{g\infty3}$●Interaction$_{g\infty4}$●*S TOP*)".

$$!IFD_g =$$

$$(Interaction_{g11} \bullet Interaction_{g12} \bullet Interaction_{g13} \bullet Interaction_{g14} \bullet STOP)$$
$$\|$$
$$(Interaction_{g21} \bullet Interaction_{g22} \bullet Interaction_{g23} \bullet Interaction_{g24} \bullet STOP)$$
$$\|$$
$$(Interaction_{g31} \bullet Interaction_{g32} \bullet Interaction_{g33} \bullet Interaction_{g34} \bullet STOP)$$
$$\ldots\ldots$$
$$\|$$
$$(Interaction_{g\infty1} \bullet Interaction_{g\infty2} \bullet Interaction_{g\infty3} \bullet Interaction_{g\infty4} \bullet STOP)$$

$!IFD_h$ describes the infinite replication of the hth interaction flow diagram, i.e. *Study_Algebra_Course* behavior, of the design view of *Kurdi University*. $!IFD_h$ is syntactically represented as "$(Interaction_{h11} \bullet Interaction_{h12} \bullet Interaction_{h13} \bullet Interaction_{h14} \bullet STOP)|$ $|(Interaction_{h21} \bullet Interaction_{h22} \bullet Interaction_{h23} \bullet Interaction_{h24} \bullet STOP)$ $\|(Interaction_{h31} \bullet Interaction_{h32} \bullet Interaction_{h33} \bullet Interaction_{h34} \bullet STOP)\ldots$ $\ldots\|(Interaction_{h\infty1} \bullet Interaction_{h\infty2} \bullet Interaction_{h\infty3} \bullet Interaction_{h\infty4} \bullet STOP)$".

$$!IFD_h =$$

$$(Interaction_{h11} \bullet Interaction_{h12} \bullet Interaction_{h13} \bullet Interaction_{h14} \bullet STOP)$$
$$\|$$
$$(Interaction_{h21} \bullet Interaction_{h22} \bullet Interaction_{h23} \bullet Interaction_{h24} \bullet STOP)$$
$$\|$$
$$(Interaction_{h31} \bullet Interaction_{h32} \bullet Interaction_{h33} \bullet Interaction_{h34} \bullet STOP)$$
$$\ldots\ldots$$
$$\|$$
$$(Interaction_{h\infty1} \bullet Interaction_{h\infty2} \bullet Interaction_{h\infty3} \bullet Interaction_{h\infty4} \bullet STOP)$$

Infinite-queue SBC process of the design view of *Kurdi University* is syntactically represented as $(!IFD_g)\|(!IFD_h)$ which equals   to "(Interaction$_{g11}$•Interaction$_{g12}$•Interaction$_{g13}$•Interaction$_{g14}$•*STOP*)| |(Interaction$_{g21}$•Interaction$_{g22}$•Interaction$_{g23}$•Interaction$_{g24}$•*STOP*) ||(Interaction$_{g31}$•Interaction$_{g32}$•Interaction$_{g33}$•Interaction$_{g34}$•*STOP*)… ….||(Interaction$_{g\infty1}$•Interaction$_{g\infty2}$•Interaction$_{g\infty3}$•Interaction$_{g\infty4}$•*S TOP*)||(Interaction$_{h11}$•Interaction$_{h12}$•Interaction$_{h13}$•Interaction$_{h14}$• *STOP*)||(Interaction$_{h21}$•Interaction$_{h22}$•Interaction$_{h23}$•Interaction$_{h24}$ •*STOP*)||(Interaction$_{h31}$•Interaction$_{h32}$•Interaction$_{h33}$•Interaction$_{h3}$ $_4$•*STOP*)…….||(Interaction$_{h\infty1}$•Interaction$_{h\infty2}$•Interaction$_{h\infty3}$•Inter action$_{h\infty4}$•*STOP*)".

Kurdi University's Design View $\overset{\text{def}}{=\!=}$

(Interaction$_{g11}$•Interaction$_{g12}$•Interaction$_{g13}$•Interaction$_{g14}$•*STOP*)
‖
(Interaction$_{g21}$•Interaction$_{g22}$•Interaction$_{g23}$•Interaction$_{g24}$•*STOP*)
‖
(Interaction$_{g31}$•Interaction$_{g32}$•Interaction$_{g33}$•Interaction$_{g34}$•*STOP*) …
‖
(Interaction$_{g\infty1}$•Interaction$_{g\infty2}$•Interaction$_{g\infty3}$•Interaction$_{g\infty4}$•*STOP*)
‖
(Interaction$_{h11}$•Interaction$_{h12}$•Interaction$_{h13}$•Interaction$_{h14}$•*STOP*)
‖
(Interaction$_{h21}$•Interaction$_{h22}$•Interaction$_{h23}$•Interaction$_{h24}$•*STOP*)
‖
(Interaction$_{h31}$•Interaction$_{h32}$•Interaction$_{h33}$•Interaction$_{h34}$•*STOP*) …
‖
(Interaction$_{h\infty1}$•Interaction$_{h\infty2}$•Interaction$_{h\infty3}$•Interaction$_{h\infty4}$•*STOP*)

Infinite-Queue SBC Process of the Structural Composition of the Design View

Structural composition of the design view of *Kurdi University* means to compose the *Mathematics_Chairman* and *Mathematics_Lecturers* components into the *Mathematics_Department* component. That is, we will rename the *Mathematics_Chairman* component to the *Mathematics_Department* component; we also will rename the *Mathematics_Lecturers* component to the *Mathematics_Department* component.

[Mathematics_Department/Mathematics_Chairman, Mathematics_Department/Mathematics_Lecturers]

We draw the infinite-queue SBC process algebra Backus-Naur Form tree of the structural composition of the design view of *Kurdi University* as follows:

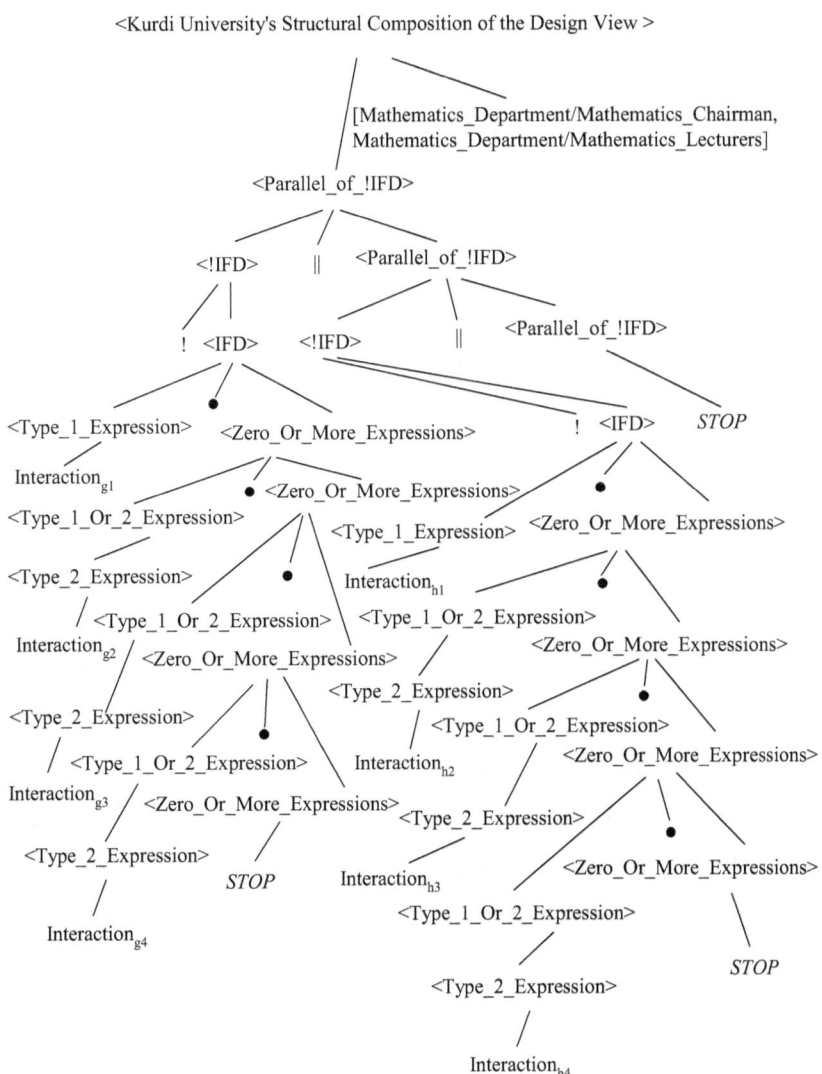

Interaction$_{i1}$=Interaction$_{g1}$[Mathematics_Department/Mathe matics_Chairman,Mathematics_Department/Mathematics_Lecturer s]=Interaction$_{c1}$ stands for the 1st interaction of the ith interaction flow diagram of the structural composition of the design view of *Kurdi University*. Interaction$_{i1}$ is a type_1 interaction which describes the *Student* actor interacts with the *University_President* component.

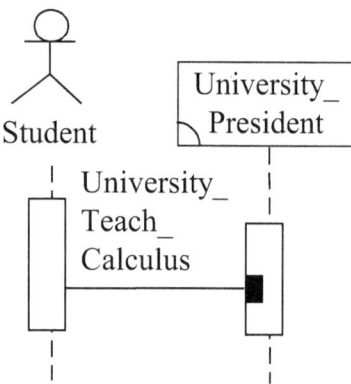

Interaction$_{i2}$=Interaction$_{g2}$[Mathematics_Department/Mathe matics_Chairman,Mathematics_Department/Mathematics_Lecturer s]=Interaction$_{c2}$ stands for the 2nd interaction of the ith interaction flow diagram of the structural composition of the design view of *Kurdi University*. Interaction$_{i2}$ is a type_2 interaction which describes the *University_President* component interacts with the *Science_Dean* component.

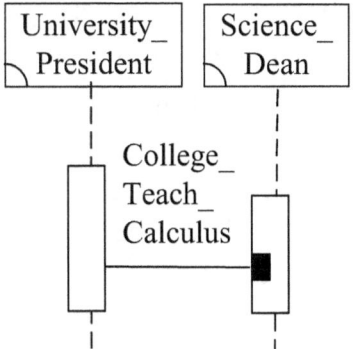

Interaction$_{i3}$=Interaction$_{g3}$[Mathematics_Department/Mathe matics_Chairman,Mathematics_Department/Mathematics_Lecturer s]=Interaction$_{c3}$ stands for the 3rd interaction of the ith interaction flow diagram of the structural composition of the design view of *Kurdi University*. Interaction$_{i3}$ is a type_2 interaction which describes the *Science_Dean* component interacts with the *Mathematics_Department* component.

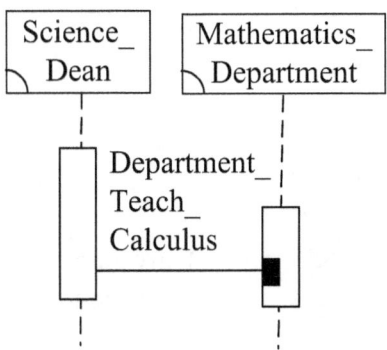

Interaction$_{i4}$=Interaction$_{g4}$[Mathematics_Department/Mathe matics_Chairman,Mathematics_Department/Mathematics_Lecturer s] stands for the 4th interaction of the ith interaction flow diagram of the structural composition of the design view of *Kurdi University*. Interaction$_{i4}$ is an internal interaction (i.e. λ) inside the *Mathematics_Department* component.

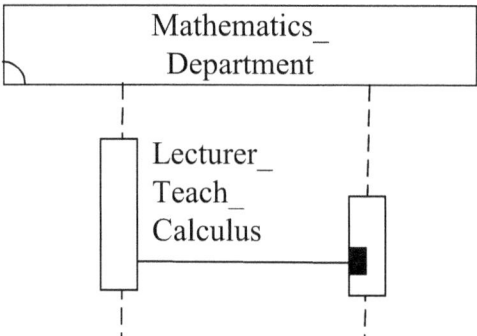

Interaction$_{j1}$=Interaction$_{h1}$[Mathematics_Department/Mathe matics_Chairman,Mathematics_Department/Mathematics_Lecturer s]=Interaction$_{d1}$ stands for the 1st interaction of the jth interaction flow diagram of the structural composition of the design view of *Kurdi University*. Interaction$_{j1}$ is a type_1 interaction which describes the *Student* actor interacts with the *University_President* component.

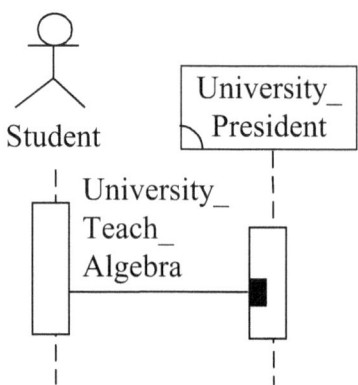

Interaction$_{j2}$=Interaction$_{h2}$[Mathematics_Department/Mathematics_Chairman,Mathematics_Department/Mathematics_Lecturers]=Interaction$_{d2}$ stands for the 2nd interaction of the jth interaction flow diagram of the structural composition of the design view of *Kurdi University*. Interaction$_{j2}$ is a type_2 interaction which describes the *University_President* component interacts with the *Science_Dean* component.

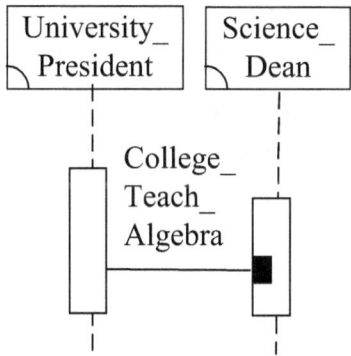

Interaction$_{j3}$=Interaction$_{h3}$[Mathematics_Department/Mathematics_Chairman,Mathematics_Department/Mathematics_Lecturers]=Interaction$_{d3}$ stands for the 3rd interaction of the jth interaction flow diagram of the structural composition of the design view of *Kurdi University*. Interaction$_{j3}$ is a type_2 interaction which describes the *Science_Dean* component interacts with the *Mathematics_Department* component.

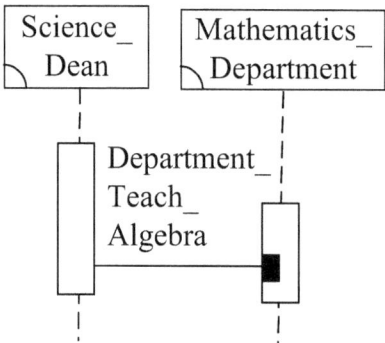

Interaction$_{j4}$=Interaction$_{h4}$[Mathematics_Department/Mathe matics_Chairman,Mathematics_Department/Mathematics_Lecturer s] stands for the 4th interaction of the jth interaction flow diagram of the structural composition of the design view of *Kurdi University*. Interaction$_{j4}$ is an internal interaction (i.e. λ) inside the *Mathematics_Department* component.

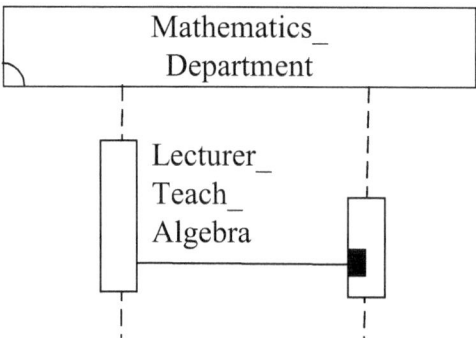

IFD$_i$ describes the ith interaction flow diagram, i.e. *Study_Calculus_Course* behavior, of the structural composition of the design view of *Kurdi University*. IFD$_i$ is syntactically represented as "Interaction$_{c1}$•Interaction$_{c2}$•Interaction$_{c3}$•λ•*STOP*".

$$\text{IFD}_i =$$

$$\text{Interaction}_{c1} \bullet \text{Interaction}_{c2} \bullet \text{Interaction}_{c3} \bullet \lambda \bullet STOP$$

IFD_j describes the jth interaction flow diagram, i.e. *Study_Algebra_Course* behavior, of the structural composition of the design view of *Kurdi University*. IFD_j is syntactically represented as "$\text{Interaction}_{d1} \bullet \text{Interaction}_{d2} \bullet \text{Interaction}_{d3} \bullet \lambda \bullet STOP$".

$$\text{IFD}_j =$$

$$\text{Interaction}_{d1} \bullet \text{Interaction}_{d2} \bullet \text{Interaction}_{d3} \bullet \lambda \bullet STOP$$

Each interaction flow diagram may replicate itself a countably infinite time. Interaction_{xyz} stands for the yth replication of the zth interaction of the xth interaction flow diagram. $!\text{IFD}_i$ describes the infinite replication of the ith interaction flow diagram, i.e. *Study_Calculus_Course* behavior, of the structural composition of the design view of *Kurdi University*. $!\text{IFD}_i$ is syntactically represented   as
"$(\text{Interaction}_{c11} \bullet \text{Interaction}_{c12} \bullet \text{Interaction}_{c13} \bullet \lambda \bullet STOP)$ $\parallel$
$(\text{Interaction}_{c21} \bullet \text{Interaction}_{c22} \bullet \text{Interaction}_{c23} \bullet \lambda \bullet STOP)$ $\parallel$
$(\text{Interaction}_{c31} \bullet \text{Interaction}_{c32} \bullet \text{Interaction}_{c33} \bullet \lambda \bullet STOP)$ .... $\parallel$
$(\text{Interaction}_{c\infty1} \bullet \text{Interaction}_{c\infty2} \bullet \text{Interaction}_{c\infty3} \bullet \lambda \bullet STOP)$".

142

$$!IFD_i =$$

$$(Interaction_{c11} \bullet Interaction_{c12} \bullet Interaction_{c13} \bullet \lambda \bullet STOP)$$
$$\|$$
$$(Interaction_{c21} \bullet Interaction_{c22} \bullet Interaction_{c23} \bullet \lambda \bullet STOP)$$
$$\|$$
$$(Interaction_{c31} \bullet Interaction_{c32} \bullet Interaction_{c33} \bullet \lambda \bullet STOP)$$
$$.......$$
$$\|$$
$$(Interaction_{c\infty1} \bullet Interaction_{c\infty2} \bullet Interaction_{c\infty3} \bullet \lambda \bullet STOP)$$

$!IFD_j$ describes the infinite replication of the jth interaction flow diagram, i.e. *Study_Algebra_Course* behavior, of the structural composition of the design view of *Kurdi University*. $!IFD_j$ is syntactically represented as "(Interaction$_{d11}$•Interaction$_{d12}$•Interaction$_{d13}$•λ•*STOP*) ‖ (Interaction$_{d21}$•Interaction$_{d22}$•Interaction$_{d23}$•λ•*STOP*) ‖ (Interaction$_{d31}$•Interaction$_{d32}$•Interaction$_{d33}$•λ•*STOP*) .... ‖ (Interaction$_{d\infty1}$•Interaction$_{d\infty2}$•Interaction$_{d\infty3}$•λ•*STOP*)".

$$!IFD_j =$$

$$(Interaction_{d11} \bullet Interaction_{d12} \bullet Interaction_{d13} \bullet \lambda \bullet STOP)$$
$$\|$$
$$(Interaction_{d21} \bullet Interaction_{d22} \bullet Interaction_{d23} \bullet \lambda \bullet STOP)$$
$$\|$$
$$(Interaction_{d31} \bullet Interaction_{d32} \bullet Interaction_{d33} \bullet \lambda \bullet STOP)$$
$$.......$$
$$\|$$
$$(Interaction_{d\infty1} \bullet Interaction_{d\infty2} \bullet Interaction_{d\infty3} \bullet \lambda \bullet STOP)$$

Infinite-queue SBC process of the structural composition of the design view of *Kurdi University* is syntactically represented as $(!IFD_i) \| (!IFD_j)$ which equals to "(Interaction$_{c11}$•Interaction$_{c12}$•Interaction$_{c13}$•λ•*STOP*) $\|$
(Interaction$_{c21}$•Interaction$_{c22}$•Interaction$_{c23}$•λ•*STOP*) $\|$
(Interaction$_{c31}$•Interaction$_{c32}$•Interaction$_{c33}$•λ•*STOP*) .... $\|$
(Interaction$_{c\infty1}$•Interaction$_{c\infty2}$•Interaction$_{c\infty3}$•λ•*STOP*) $\|$
(Interaction$_{d11}$•Interaction$_{d12}$•Interaction$_{d13}$•λ•*STOP*) $\|$
(Interaction$_{d21}$•Interaction$_{d22}$•Interaction$_{d23}$•λ•*STOP*) $\|$
(Interaction$_{d31}$•Interaction$_{d32}$•Interaction$_{d33}$•λ•*STOP*) .... $\|$
(Interaction$_{d\infty1}$•Interaction$_{d\infty2}$•Interaction$_{d\infty3}$•λ•*STOP*)".

Kurdi University's Structural Composition of the Design View $\overset{def}{=\!=}$

(Interaction$_{c11}$•Interaction$_{c12}$•Interaction$_{c13}$•λ•*STOP*)
$\quad\|$
(Interaction$_{c21}$•Interaction$_{c22}$•Interaction$_{c23}$•λ•*STOP*)
$\quad\|$
(Interaction$_{c31}$•Interaction$_{c32}$•Interaction$_{c33}$•λ•*STOP*) ....
$\quad\|$
(Interaction$_{c\infty1}$•Interaction$_{c\infty2}$•Interaction$_{c\infty3}$•λ•*STOP*)
$\|$
(Interaction$_{d11}$•Interaction$_{d12}$•Interaction$_{d13}$•λ•*STOP*)
$\quad\|$
(Interaction$_{d21}$•Interaction$_{d22}$•Interaction$_{d23}$•λ•*STOP*)
$\quad\|$
(Interaction$_{d31}$•Interaction$_{d32}$•Interaction$_{d33}$•λ•*STOP*) ....
$\quad\|$
(Interaction$_{d\infty1}$•Interaction$_{d\infty2}$•Interaction$_{d\infty3}$•λ•*STOP*)

Observation Congruence of "the Analysis View" and "the Structural Composition of the Design View"

We syntactically represent infinite-queue SBC process P_{01} as $\text{IFD}_c \| \text{IFD}_d$ which equals to $(\text{Interaction}_{c1} \bullet \text{Interaction}_{c2} \bullet \text{Interaction}_{c3} \bullet STOP) \| (\text{Interaction}_{d1} \bullet \text{Interaction}_{d2} \bullet \text{Interaction}_{d3} \bullet STOP)$. We also syntactically represent infinite-queue SBC processes P_{02} as $(\text{Interaction}_{c2} \bullet \text{Interaction}_{c3} \bullet STOP) \| (\text{Interaction}_{d1} \bullet \text{Interaction}_{d2} \bullet \text{Interaction}_{d3} \bullet STOP)$, P_{03} as $(\text{Interaction}_{c3} \bullet STOP) \| (\text{Interaction}_{d1} \bullet \text{Interaction}_{d2} \bullet \text{Interaction}_{d3} \bullet STOP)$, P_{04} as $(\text{Interaction}_{d1} \bullet \text{Interaction}_{d2} \bullet \text{Interaction}_{d3} \bullet STOP)$, P_{05} as $(\text{Interaction}_{c1} \bullet \text{Interaction}_{c2} \bullet \text{Interaction}_{c3} \bullet STOP) \| (\text{Interaction}_{d2} \bullet \text{Interaction}_{d3} \bullet STOP)$, P_{06} as $(\text{Interaction}_{c2} \bullet \text{Interaction}_{c3} \bullet STOP) \| (\text{Interaction}_{d2} \bullet \text{Interaction}_{d3} \bullet STOP)$, P_{07} as $(\text{Interaction}_{c3} \bullet STOP) \| (\text{Interaction}_{d2} \bullet \text{Interaction}_{d3} \bullet STOP)$, P_{08} as $(\text{Interaction}_{d2} \bullet \text{Interaction}_{d3} \bullet STOP)$, P_{09} as $(\text{Interaction}_{c1} \bullet \text{Interaction}_{c2} \bullet \text{Interaction}_{c3} \bullet STOP) \| (\text{Interaction}_{d3} \bullet STOP)$, P_{10} as $(\text{Interaction}_{c2} \bullet \text{Interaction}_{c3} \bullet STOP) \| (\text{Interaction}_{d3} \bullet STOP)$, P_{11} as $(\text{Interaction}_{c3} \bullet STOP) \| (\text{Interaction}_{d3} \bullet STOP)$, P_{12} as

(Interaction$_{d3}$•*STOP*), P_{13} as (Interaction$_{c1}$•Interaction$_{c2}$•Interaction$_{c3}$•*STOP*), P_{14} as (Interaction$_{c2}$•Interaction$_{c3}$•*STOP*), P_{15} as (Interaction$_{c3}$•*STOP*), P_{16} as *STOP*, respectively.

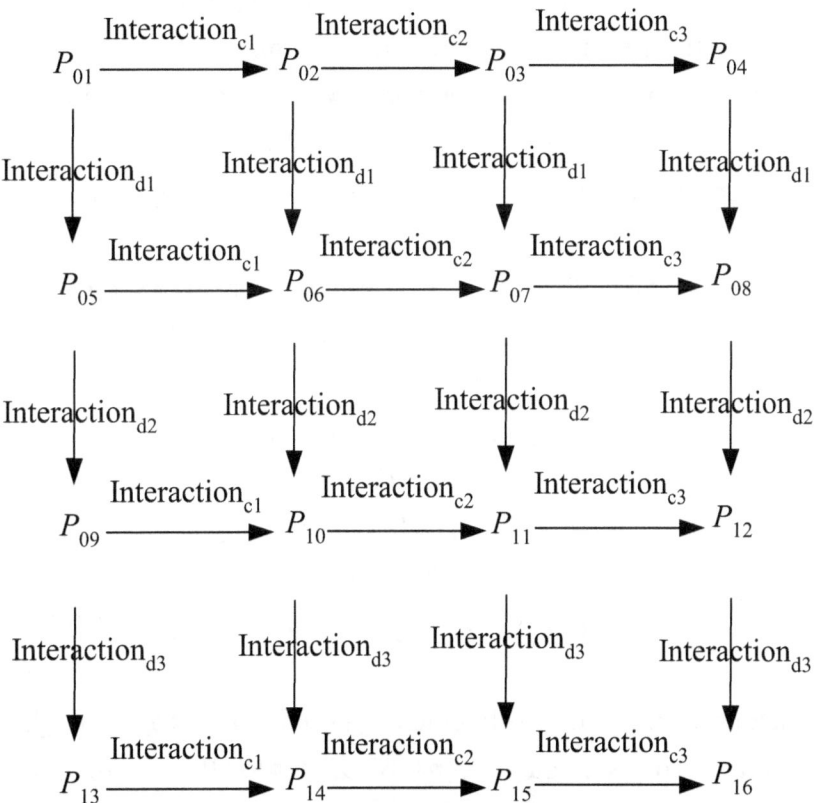

We syntactically represent process Q_{01} as IFD$_i$‖IFD$_j$ which equals to (Interaction$_{c1}$•Interaction$_{c2}$•Interaction$_{c3}$•λ•*STOP*)‖(Interaction$_{d1}$•Interaction$_{d2}$•Interaction$_{d3}$•λ•*STOP*). We also syntactically represent processes Q_{02} as ($\text{Interaction}_{c2} \bullet \text{Interaction}_{c3} \bullet \lambda \bullet STOP$)$\|$($\text{Interaction}_{d1} \bullet \text{Interaction}_{d2} \bullet \text{Interaction}_{d3} \bullet \lambda \bullet STOP$), Q_{03} as ($\text{Interaction}_{c3} \bullet \lambda \bullet STOP$)$\|$($\text{Interaction}_{d1} \bullet \text{Interaction}_{d2} \bullet \text{Interaction}_{d3} \bullet \lambda \bullet STOP$), Q_{04} as ($\lambda \bullet STOP\|\text{Interaction}_{d1} \bullet \text{Interaction}_{d2} \bullet \text{Interaction}_{d3} \bullet \lambda \bullet STOP$), Q_{05} as ($\text{Interaction}_{d1} \bullet \text{Interaction}_{d2} \bullet \text{Interaction}_{d3} \bullet \lambda \bullet STOP$), Q_{06} as ($\text{Interaction}_{c1} \bullet \text{Interaction}_{c2} \bullet \text{Interaction}_{c3} \bullet \lambda \bullet STOP$)$\|$($\text{Interaction}_{d2} \bullet \text{Interaction}_{d3} \bullet \lambda \bullet STOP$), Q_{07} as ($\text{Interaction}_{c2} \bullet \text{Interaction}_{c3} \bullet \lambda \bullet STOP$)$\|$($\text{Interaction}_{d2} \bullet \text{Interaction}_{d3} \bullet \lambda \bullet STOP$), Q_{08} as ($\text{Interaction}_{c3} \bullet \lambda \bullet STOP$)$\|$($\text{Interaction}_{d2} \bullet \text{Interaction}_{d3} \bullet \lambda \bullet STOP$), Q_{09} as ($\lambda \bullet STOP\|\text{Interaction}_{d2} \bullet \text{Interaction}_{d3} \bullet \lambda \bullet STOP$), Q_{10} as ($\text{Interaction}_{d2} \bullet \text{Interaction}_{d3} \bullet \lambda \bullet STOP$), Q_{11} as ($\text{Interaction}_{c1} \bullet \text{Interaction}_{c2} \bullet \text{Interaction}_{c3} \bullet \lambda \bullet STOP$)$\|$($\text{Interaction}_{d3} \bullet \lambda \bullet STOP$), Q_{12} as ($\text{Interaction}_{c2} \bullet \text{Interaction}_{c3} \bullet \lambda \bullet STOP$)$\|$($\text{Interaction}_{d3} \bullet \lambda \bullet STOP$), Q_{13} as ($\text{Interaction}_{c3} \bullet \lambda \bullet STOP$)$\|$($\text{Interaction}_{d3} \bullet \lambda \bullet STOP$), Q_{14} as ($\lambda \bullet STOP\|\text{Interaction}_{d3} \bullet \lambda \bullet STOP$), Q_{15} as ($\text{Interaction}_{d3} \bullet \lambda \bullet STOP$), Q_{16} as ($\text{Interaction}_{c1} \bullet \text{Interaction}_{c2} \bullet \text{Interaction}_{c3} \bullet \lambda \bullet STOP\|(\lambda \bullet STOP)$, Q_{17} as ($\text{Interaction}_{c2} \bullet \text{Interaction}_{c3} \bullet \lambda \bullet STOP\|(\lambda \bullet STOP)$, Q_{18} as ($\text{Interaction}_{c3} \bullet \lambda \bullet STOP\|(\lambda \bullet STOP)$, Q_{19} as ($\lambda \bullet STOP\|(\lambda \bullet STOP)$, Q_{20} as ($\lambda \bullet STOP$), Q_{21} as ($\text{Interaction}_{c1} \bullet \text{Interaction}_{c2} \bullet \text{Interaction}_{c3} \bullet \lambda \bullet STOP$), Q_{22} as ($\text{Interaction}_{c2} \bullet \text{Interaction}_{c3} \bullet \lambda \bullet STOP$), Q_{23} as ($\text{Interaction}_{c3} \bullet \lambda \bullet STOP$), Q_{24} as ($\lambda \bullet STOP$), Q_{25} as $STOP$, respectively.

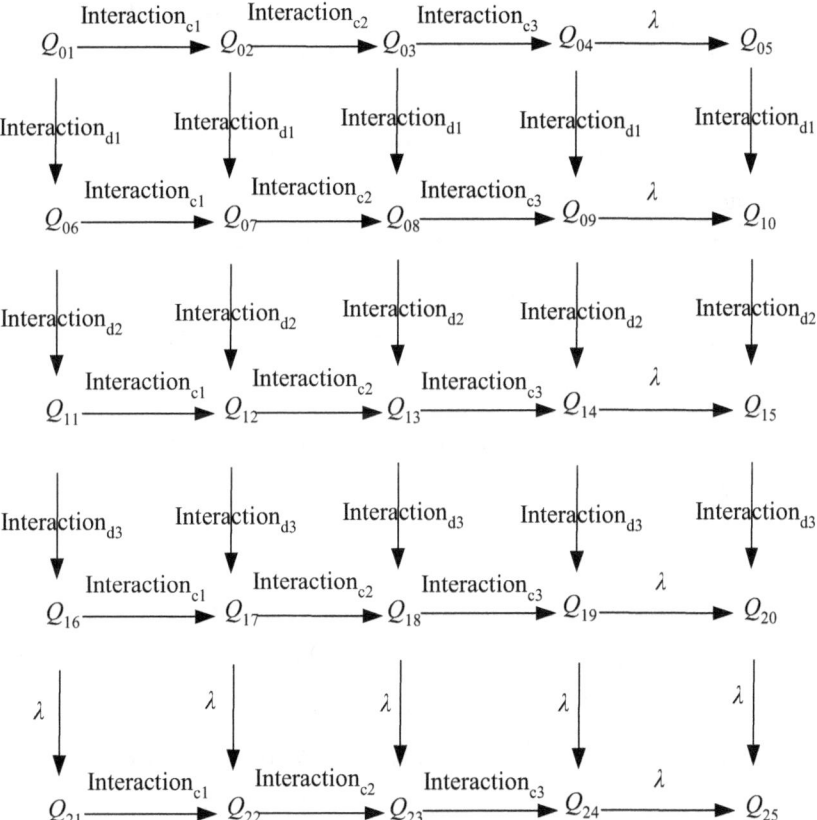

We can easily verify that $S = \{(P_{01}, Q_{01}), (P_{02}, Q_{02}), (P_{03}, Q_{03}), (P_{04}, Q_{04}), (P_{04}, Q_{05}), (P_{05}, Q_{06}), (P_{06}, Q_{07}), (P_{07}, Q_{08}), (P_{08}, Q_{09}), (P_{08}, Q_{10}), (P_{09}, Q_{11}), (P_{10}, Q_{12}), (P_{11}, Q_{13}), (P_{12}, Q_{14}), (P_{12}, Q_{15}), (P_{13}, Q_{16}), (P_{13}, Q_{21}), (P_{14}, Q_{17}), (P_{14}, Q_{22}), (P_{15}, Q_{18}), (P_{15}, Q_{23}), (P_{16}, Q_{19}), (P_{16}, Q_{20}), (P_{16}, Q_{24}), (P_{16}, Q_{25})\}$ is a bisimulation.

Using the S bisimulation, we then are able to verify that P_{01} and Q_{01} are observation congruent because (1) $P_{01} \xrightarrow{\text{Interaction}_{c1}} P_{02}$, then we have Q_{02} that $Q_{01} \xRightarrow{\text{Interaction}_{c1}} Q_{02}$ and $P_{02} \stackrel{\approx}{\sim} Q_{02}$, and (2) $P_{01} \xrightarrow{\text{Interaction}_{d1}} P_{05}$, then we have Q_{06} that $Q_{01} \xRightarrow{\text{Interaction}_{d1}} Q_{06}$ and $P_{05} \stackrel{\approx}{\sim} Q_{06}$, and (3) $Q_{01} \xrightarrow{\text{Interaction}_{c1}} Q_{02}$, then we have P_{02} that $P_{01} \xRightarrow{\text{Interaction}_{c1}} P_{02}$ and $P_{02} \stackrel{\approx}{\sim} Q_{02}$, and (4) $Q_{01} \xrightarrow{\text{Interaction}_{d1}} Q_{06}$, then we have P_{05} that $P_{01} \xRightarrow{\text{Interaction}_{d1}} P_{05}$ and $P_{05} \stackrel{\approx}{\sim} Q_{06}$.

$P_{01} = Q_{01}$ means $IFD_c \| IFD_d = IFD_i \| IFD_j$, and $IFD_c \| IFD_d = IFD_i \| IFD_j$ means $!(IFD_c \| IFD_d) = !(IFD_i \| IFD_j)$, and $!(IFD_c \| IFD_d) = !(IFD_i \| IFD_j)$ means $(!IFD_c) \| (!IFD_d) = (!IFD_i) \| (!IFD_j)$. So, there is observation congruence of "the analysis view of *Kurdi University*" and "the structural composition of the design view of *Kurdi University*".

Conclusively, the design view of *Kurdi University* is one level down structural decomposition (with observation congruence verification) of the analysis view of *Kurdi University*.

Infinite-Queue SBC Process of the Implementation View

We draw the Architecture Hierarchy Diagram (AHD) of the implementation view of *Kurdi University* as follows:

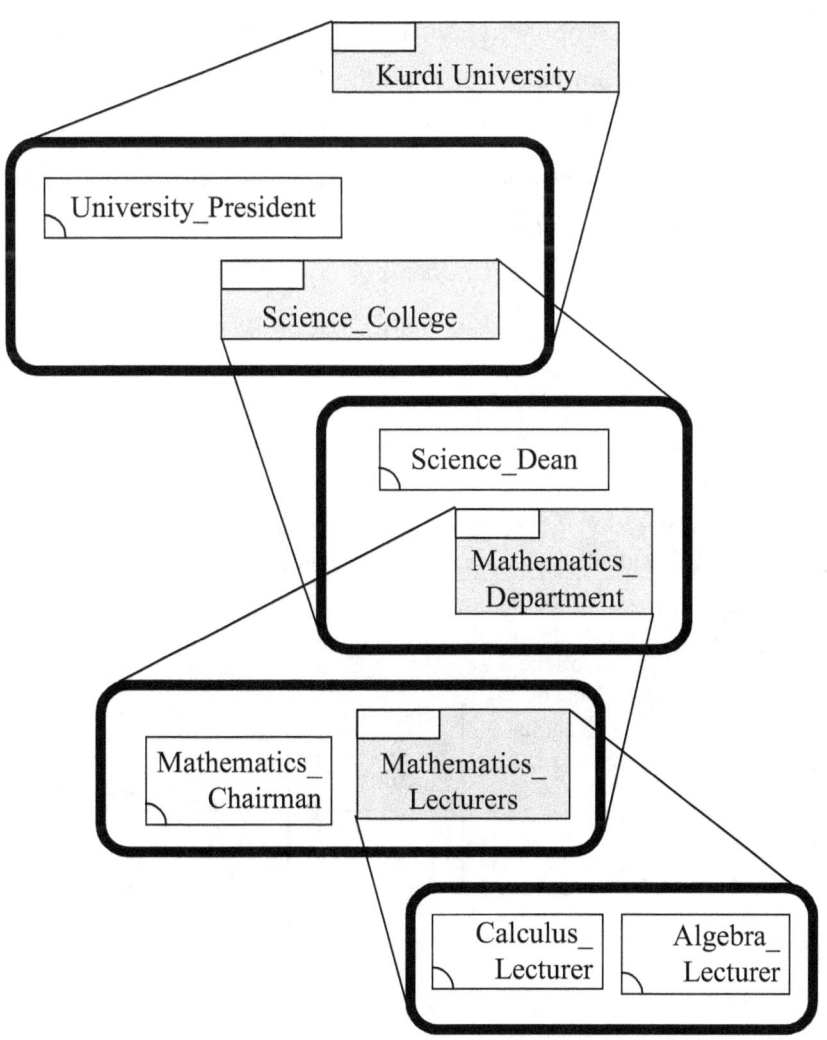

The overall behavior of the implementation view of *Kurdi University* includes two behaviors: *Study_Calculus_Course* and *Study_Algebra_Course*. Each of them is described by an individual IFD.

An IFD of the *Study_Calculus_Course* behavior is shown below. First, actor *Student* interacts with the *University_President* component through the *University_Teach_Calculus* operation call interaction. Next, component *University_President* interacts with the *Science_Dean* component through the *College_Teach_Calculus* operation call interaction. Continuingly, component *Science_Dean* interacts with the *Mathematics_Chairman* component through the *Department_Teach_Calculus* operation call interaction. Finally, component *Mathematics_Chairman* interacts with the *Calculus_Lecturer* component through the *Lecturer_Teach_Calculus* operation call interaction.

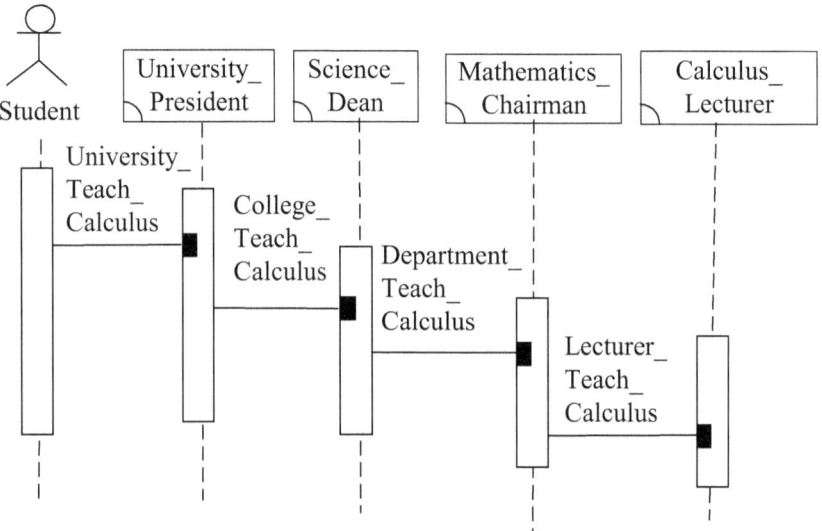

An IFD of the *Study_Algebra_Course* behavior is shown below. First, actor *Student* interacts with the *University_President* component through the *University_Teach_Algebra* operation call interaction. Next, component *University_President* interacts with the *Science_Dean* component through the *College_Teach_Algebra* operation call interaction. Continuingly, component *Science_Dean* interacts with the *Mathematics_Chairman* component through the *Department_Teach_Algebra* operation call interaction. Finally, component *Mathematics_Chairman* interacts with the *Algebra_Lecturer* component through the *Lecturer_Teach_Algebra* operation call interaction.

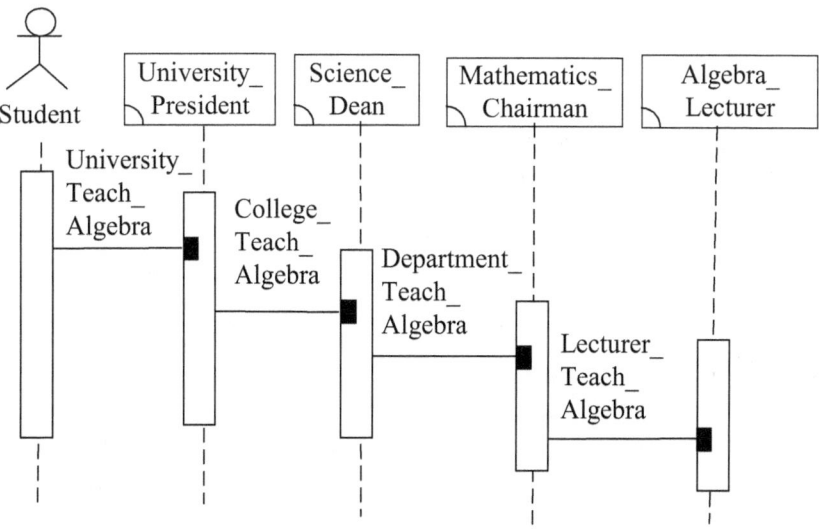

We draw the infinite-queue SBC process algebra Backus-Naur Form tree of the implementation view of *Kurdi University* as follows:

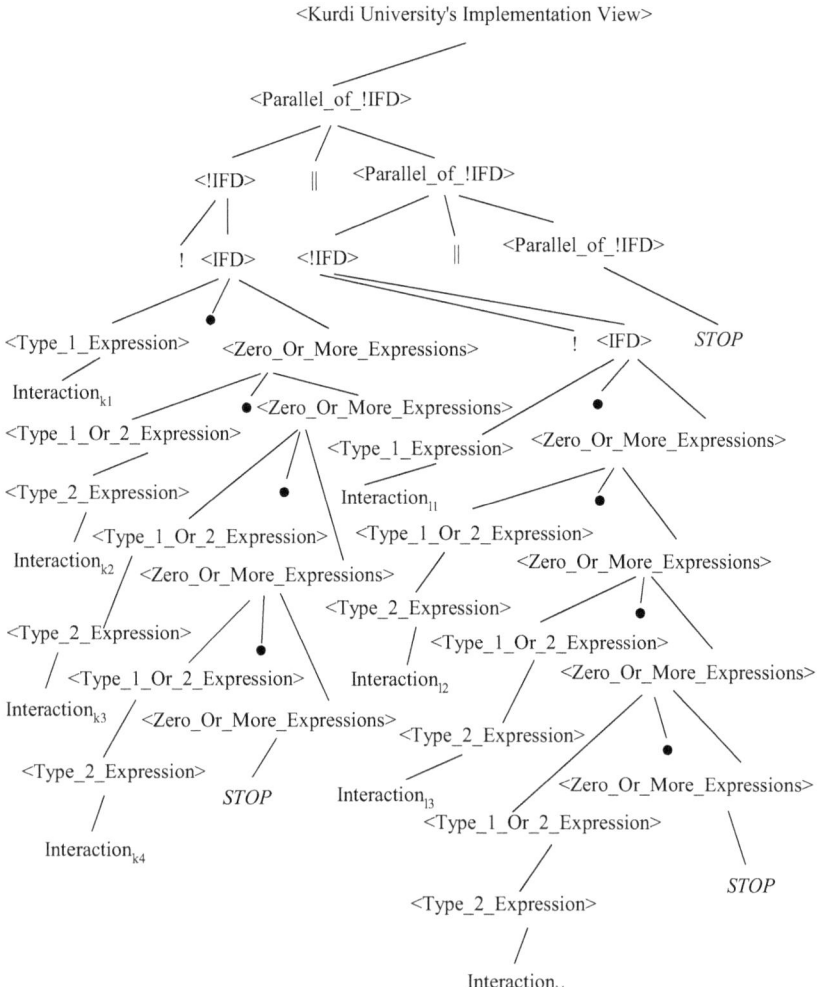

Interaction$_{k1}$ stands for the 1st interaction of the kth interaction flow diagram of the implementation view of *Kurdi University*. Interaction$_{k1}$ is a type_1 interaction which describes the *Student* actor interacts with the *University_President* component.

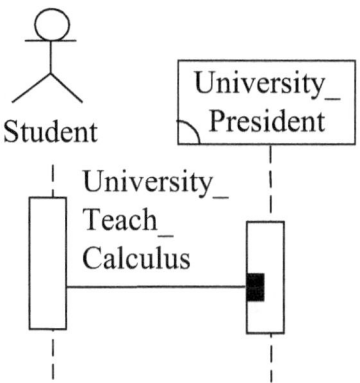

Interaction$_{k2}$ stands for the 2nd interaction of the kth interaction flow diagram of the implementation view of *Kurdi University*. Interaction$_{k2}$ is a type_2 interaction which describes the *University_President* component interacts with the *Science_Dean* component.

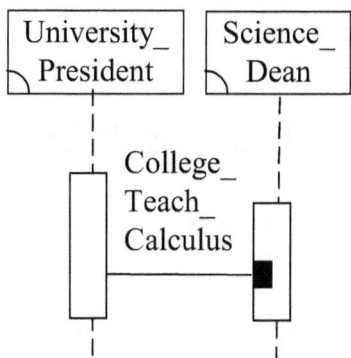

Interaction$_{k3}$ stands for the 3rd interaction of the kth interaction flow diagram of the implementation view of *Kurdi University*. Interaction$_{k3}$ is a type_2 interaction which describes the *Science_Dean* component interacts with the *Mathematics_Chairman* component.

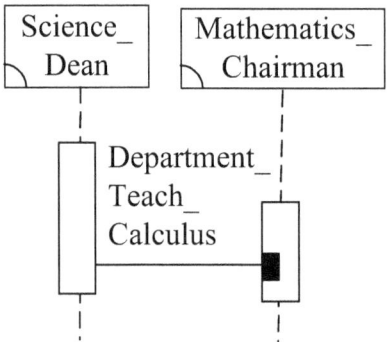

Interaction$_{k4}$ stands for the 4th interaction of the kth interaction flow diagram of the implementation view of *Kurdi University*. Interaction$_{k4}$ is a type_2 interaction which describes the *Mathematics_Chairman* component interacts with the *Calculus_Lecturer* component.

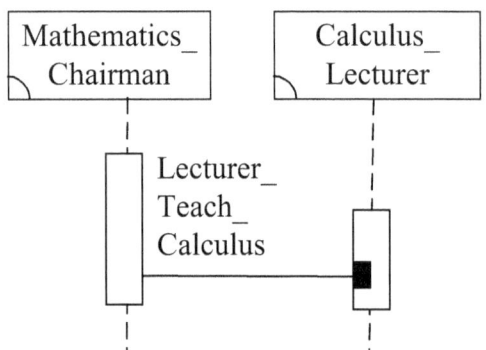

Interaction$_{11}$ stands for the 1st interaction of the lth interaction flow diagram of the implementation view of *Kurdi University*. Interaction$_{11}$ is a type_1 interaction which describes the *Student* actor interacts with the *University_President* component.

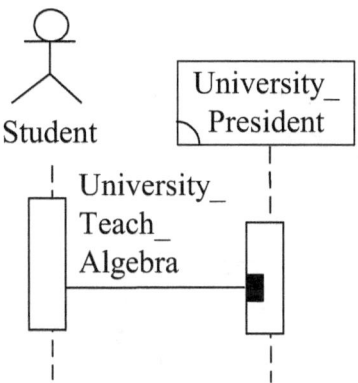

Interaction$_{12}$ stands for the 2nd interaction of the lth interaction flow diagram of the implementation view of *Kurdi University*. Interaction$_{12}$ is a type_2 interaction which describes the *University_President* component interacts with the *Science_Dean* component.

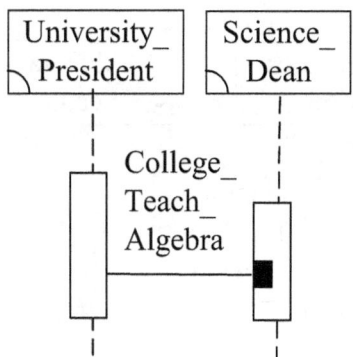

Interaction$_{l3}$ stands for the 3rd interaction of the lth interaction flow diagram of the implementation view of *Kurdi University*. Interaction$_{l3}$ is a type_2 interaction which describes the *Science_Dean* component interacts with the *Mathematics_Chairman* component.

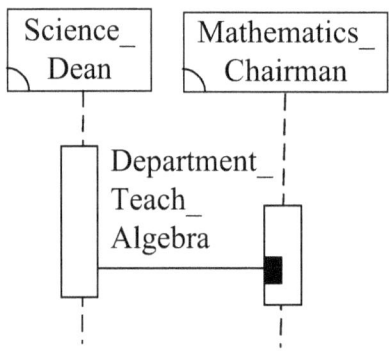

Interaction$_{l4}$ stands for the 4th interaction of the lth interaction flow diagram of the implementation view of *Kurdi University*. Interaction$_{l4}$ is a type_2 interaction which describes the *Mathematics_Chairman* component interacts with the *Algebra_Lecturer* component.

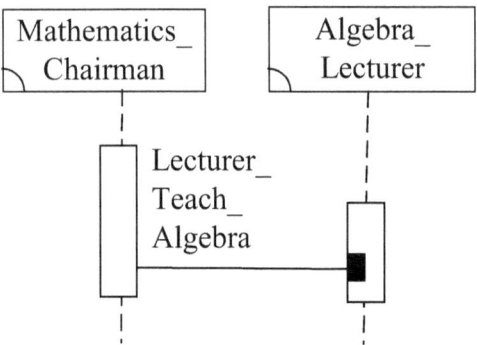

IFD_k describes the kth interaction flow diagram, i.e. *Study_Calculus_Course* behavior, of the implementation view of *Kurdi University*. IFD_k is syntactically represented as "$Interaction_{k1} \bullet Interaction_{k2} \bullet Interaction_{k3} \bullet Interaction_{k4} \bullet STOP$".

$IFD_k =$

$Interaction_{k1} \bullet Interaction_{k2} \bullet Interaction_{k3} \bullet Interaction_{k4} \bullet STOP$

IFD_l describes the lth interaction flow diagram, i.e. *Study_Algebra_Course* behavior, of the implementation view of *Kurdi University*. IFD_l is syntactically represented as "$Interaction_{l1} \bullet Interaction_{l2} \bullet Interaction_{l3} \bullet Interaction_{l4} \bullet STOP$".

$IFD_l =$

$Interaction_{l1} \bullet Interaction_{l2} \bullet Interaction_{l3} \bullet Interaction_{l4} \bullet STOP$

Each interaction flow diagram may replicate itself a countably infinite time. $Interaction_{xyz}$ stands for the yth replication of the zth interaction of the xth interaction flow diagram. $!IFD_k$ describes the infinite replication of the kth interaction flow diagram, i.e. *Study_Calculus_Course* behavior, of the implementation view of *Kurdi University*. $!IFD_k$ is syntactically represented as "$(Interaction_{k11} \bullet Interaction_{k12} \bullet Interaction_{k13} \bullet Interaction_{k14} \bullet STOP)|$ $|(Interaction_{k21} \bullet Interaction_{k22} \bullet Interaction_{k23} \bullet Interaction_{k24} \bullet STOP)$ $||(Interaction_{k31} \bullet Interaction_{k32} \bullet Interaction_{k33} \bullet Interaction_{k34} \bullet STOP)...$ $....||(Interaction_{k\infty1} \bullet Interaction_{k\infty2} \bullet Interaction_{k\infty3} \bullet Interaction_{k\infty4} \bullet S TOP)$".

$!IFD_k =$

$(Interaction_{k11} \bullet Interaction_{k12} \bullet Interaction_{k13} \bullet Interaction_{k14} \bullet STOP)$

$\|$

$(Interaction_{k21} \bullet Interaction_{k22} \bullet Interaction_{k23} \bullet Interaction_{k24} \bullet STOP)$

$\|$

$(Interaction_{k31} \bullet Interaction_{k32} \bullet Interaction_{k33} \bullet Interaction_{k34} \bullet STOP)$

....

$\|$

$(Interaction_{k\infty1} \bullet Interaction_{k\infty2} \bullet Interaction_{k\infty3} \bullet Interaction_{k\infty4} \bullet STOP)$

$!IFD_l$ describes the infinite replication of the lth interaction flow diagram, i.e. *Study_Algebra_Course* behavior, of the implementation view of *Kurdi University*. $!IFD_l$ is syntactically represented   as "$(Interaction_{l11} \bullet Interaction_{l12} \bullet Interaction_{l13} \bullet Interaction_{l14} \bullet STOP)\|$ $(Interaction_{l21} \bullet Interaction_{l22} \bullet Interaction_{l23} \bullet Interaction_{l24} \bullet STOP)\|($ $Interaction_{l31} \bullet Interaction_{l32} \bullet Interaction_{l33} \bullet Interaction_{l34} \bullet STOP)$....... $\|(Interaction_{l\infty1} \bullet Interaction_{l\infty2} \bullet Interaction_{l\infty3} \bullet Interaction_{l\infty4} \bullet STOP)$"

.

$!IFD_l =$

$(Interaction_{l11} \bullet Interaction_{l12} \bullet Interaction_{l13} \bullet Interaction_{l14} \bullet STOP)$

$\|$

$(Interaction_{l21} \bullet Interaction_{l22} \bullet Interaction_{l23} \bullet Interaction_{l24} \bullet STOP)$

$\|$

$(Interaction_{l31} \bullet Interaction_{l32} \bullet Interaction_{l33} \bullet Interaction_{l34} \bullet STOP)$

....

$\|$

$(Interaction_{l\infty1} \bullet Interaction_{l\infty2} \bullet Interaction_{l\infty3} \bullet Interaction_{l\infty4} \bullet STOP)$

Infinite-queue SBC process of the implementation view of *Kurdi University* is syntactically represented as $(!IFD_k)\|(!IFD_l)$

which   equals   to

"$(\text{Interaction}_{k11} \bullet \text{Interaction}_{k12} \bullet \text{Interaction}_{k13} \bullet \text{Interaction}_{k14} \bullet STOP)|$
$|(\text{Interaction}_{k21} \bullet \text{Interaction}_{k22} \bullet \text{Interaction}_{k23} \bullet \text{Interaction}_{k24} \bullet STOP)$
$\|(\text{Interaction}_{k31} \bullet \text{Interaction}_{k32} \bullet \text{Interaction}_{k33} \bullet \text{Interaction}_{k34} \bullet STOP)\ldots$
$\ldots\|(\text{Interaction}_{k\infty1} \bullet \text{Interaction}_{k\infty2} \bullet \text{Interaction}_{k\infty3} \bullet \text{Interaction}_{k\infty4} \bullet S$
$TOP)\|(\text{Interaction}_{l11} \bullet \text{Interaction}_{l12} \bullet \text{Interaction}_{l13} \bullet \text{Interaction}_{l14} \bullet S$
$TOP)\|(\text{Interaction}_{l21} \bullet \text{Interaction}_{l22} \bullet \text{Interaction}_{l23} \bullet \text{Interaction}_{l24} \bullet S$
$TOP)\|(\text{Interaction}_{l31} \bullet \text{Interaction}_{l32} \bullet \text{Interaction}_{l33} \bullet \text{Interaction}_{l34} \bullet S$
$TOP)\ldots\ldots\|(\text{Interaction}_{l\infty1} \bullet \text{Interaction}_{l\infty2} \bullet \text{Interaction}_{l\infty3} \bullet \text{Interactio}$
$n_{l\infty4} \bullet STOP)$".

Kurdi University's Implementation View $\overset{\text{def}}{=\!=}$

$(\text{Interaction}_{k11} \bullet \text{Interaction}_{k12} \bullet \text{Interaction}_{k13} \bullet \text{Interaction}_{k14} \bullet STOP)$
$\|$
$(\text{Interaction}_{k21} \bullet \text{Interaction}_{k22} \bullet \text{Interaction}_{k23} \bullet \text{Interaction}_{k24} \bullet STOP)$
$\|$
$(\text{Interaction}_{k31} \bullet \text{Interaction}_{k32} \bullet \text{Interaction}_{k33} \bullet \text{Interaction}_{k34} \bullet STOP) \ldots$
$\|$
$(\text{Interaction}_{k\infty1} \bullet \text{Interaction}_{k\infty2} \bullet \text{Interaction}_{k\infty3} \bullet \text{Interaction}_{k\infty4} \bullet STOP)$
$\|$
$(\text{Interaction}_{l11} \bullet \text{Interaction}_{l12} \bullet \text{Interaction}_{l13} \bullet \text{Interaction}_{l14} \bullet STOP)$
$\|$
$(\text{Interaction}_{l21} \bullet \text{Interaction}_{l22} \bullet \text{Interaction}_{l23} \bullet \text{Interaction}_{l24} \bullet STOP)$
$\|$
$(\text{Interaction}_{l31} \bullet \text{Interaction}_{l32} \bullet \text{Interaction}_{l33} \bullet \text{Interaction}_{l34} \bullet STOP) \ldots$
$\|$
$(\text{Interaction}_{l\infty1} \bullet \text{Interaction}_{l\infty2} \bullet \text{Interaction}_{l\infty3} \bullet \text{Interaction}_{l\infty4} \bullet STOP)$

Infinite-Queue SBC Process of the Structural Composition of the Implementation View

Structural composition of the implementation view of *Kurdi University* means to compose the *Calculus_Lecturer* and *Algebra_Lecturer* components into the *Mathematics_Lecturers* component. That is, we will rename the *Calculus_Lecturer* component to the *Mathematics_Lecturers* component; we also will rename the *Algebra_Lecturer* component to the *Mathematics_Lecturers* component.

```
[Mathematics_Lecturers/Calculus_Lecturer,
Mathematics_Lecturers/Algebra_Lecturer]
```

We draw the Backus-Naur Form tree of the structural composition of the implementation view of *Kurdi University* as follows:

<Kurdi University's Structural Composition of the Implementation View >

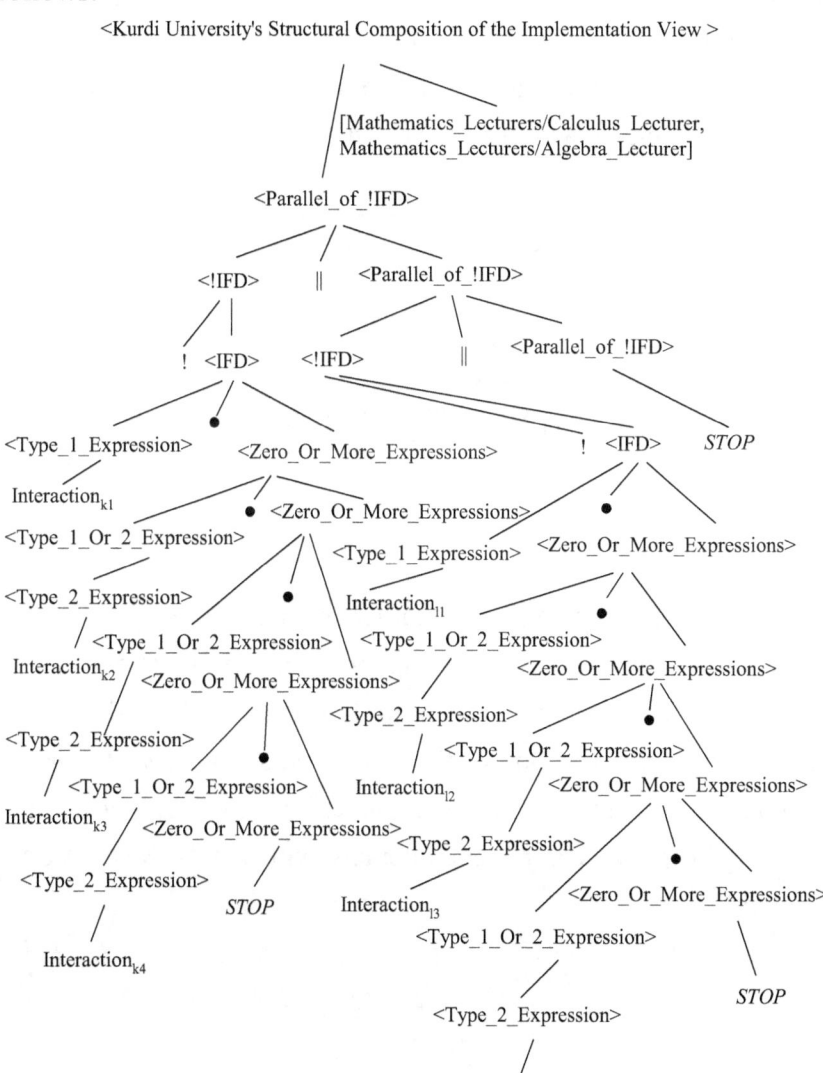

Interaction$_{m1}$=Interaction$_{k1}$[Mathematics_Lecturers/Calculus _Lecturer,Mathematics_Lecturers/Algebra_Lecturer]=Interaction$_{g1}$ stands for the 1st interaction of the mth interaction flow diagram of the structural composition of the implementation view of *Kurdi University*. Interaction$_{m1}$ is a type_1 interaction which describes the *Student* actor interacts with the *University_President* component.

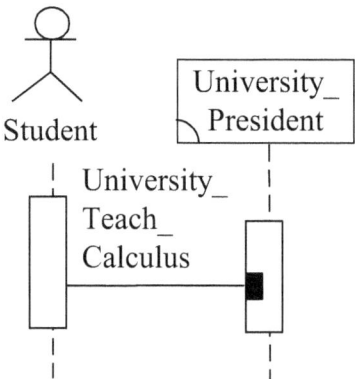

Interaction$_{m2}$=Interaction$_{k2}$[Mathematics_Lecturers/Calculus _Lecturer,Mathematics_Lecturers/Algebra_Lecturer]=Interaction$_{g2}$ stands for the 2nd interaction of the mth interaction flow diagram of the structural composition of the implementation view of *Kurdi University*. Interaction$_{m2}$ is a type_2 interaction which describes the *University_President* component interacts with the *Science_Dean* component.

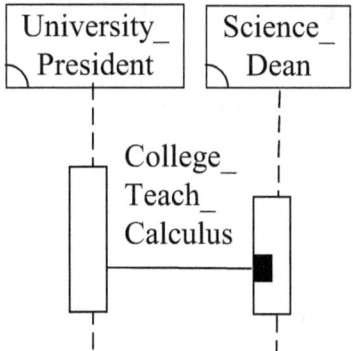

Interaction$_{m3}$=Interaction$_{k3}$[Mathematics_Lecturers/Calculus _Lecturer,Mathematics_Lecturers/Algebra_Lecturer]=Interaction$_{g3}$ stands for the 3rd interaction of the mth interaction flow diagram of the structural composition of the implementation view of *Kurdi University*. Interaction$_{m3}$ is a type_2 interaction which describes the *Science_Dean* component interacts with the *Mathematics_Chairman* component.

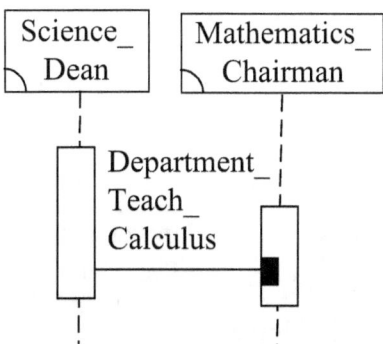

Interaction$_{m4}$=Interaction$_{k4}$[Mathematics_Lecturers/Calculus _Lecturer,Mathematics_Lecturers/Algebra_Lecturer]=Interaction$_{g4}$ stands for the 4th interaction of the mth interaction flow diagram of the structural composition of the implementation view of *Kurdi University*. Interaction$_{m4}$ is a type_2 interaction which describes the *Mathematics_Chairman* component interacts with the *Mathematics_Lecturers* component.

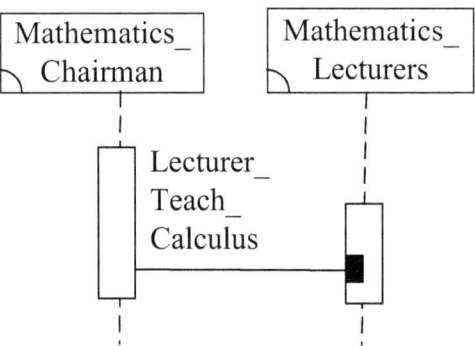

Interaction$_{n1}$=Interaction$_{l1}$[Mathematics_Lecturers/Calculus_Lecturer,Mathematics_Lecturers/Algebra_Lecturer]=Interaction$_{h1}$ stands for the 1st interaction of the nth interaction flow diagram of the structural composition of the implementation view of *Kurdi University*. Interaction$_{n1}$ is a type_1 interaction which describes the *Student* actor interacts with the *University_President* component.

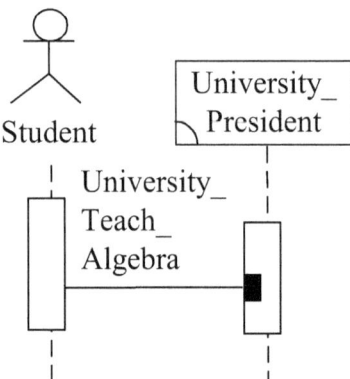

Interaction$_{n2}$=Interaction$_{i2}$[Mathematics_Lecturers/Calculus_
Lecturer,Mathematics_Lecturers/Algebra_Lecturer]=Interaction$_{h2}$
stands for the 2nd interaction of the nth interaction flow diagram of
the structural composition of the implementation view of *Kurdi
University*. Interaction$_{n2}$ is a type_2 interaction which describes the
University_President component interacts with the *Science_Dean*
component.

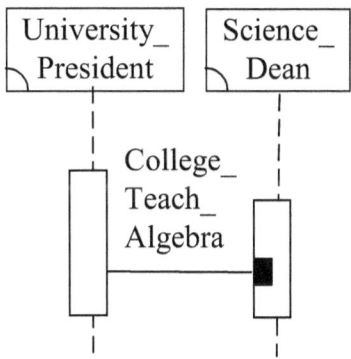

Interaction$_{n3}$=Interaction$_{i3}$[Mathematics_Lecturers/Calculus_
Lecturer,Mathematics_Lecturers/Algebra_Lecturer]=Interaction$_{h3}$
stands for the 3rd interaction of the nth interaction flow diagram of
the structural composition of the implementation view of *Kurdi
University*. Interaction$_{n3}$ is a type_2 interaction which describes the
Science_Dean component interacts with the
Mathematics_Chairman component.

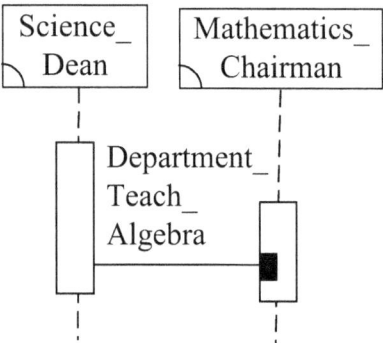

Interaction$_{n4}$=Interaction$_{14}$[Mathematics_Lecturers/Calculus_
Lecturer,Mathematics_Lecturers/Algebra_Lecturer]=Interaction$_{h4}$
stands for the 4th interaction of the nth interaction flow diagram of
the structural composition of the implementation view of *Kurdi
University*. Interaction$_{n4}$ is a type_2 interaction which describes the
Mathematics_Chairman component interacts with the
Mathematics_Lecturers component.

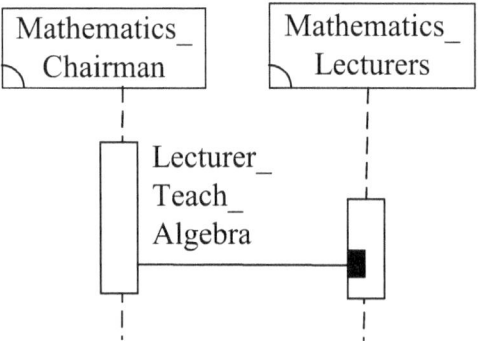

IFD$_m$ describes the mth interaction flow diagram, i.e.
Study_Calculus_Course behavior, of the structural composition of
the implementation view of *Kurdi University*. IFD$_m$ is syntactically
represented   as
"Interaction$_{g1}$•Interaction$_{g2}$•Interaction$_{g3}$•Interaction$_{g4}$•*STOP*".

$$\text{IFD}_m =$$

$$\text{Interaction}_{g1} \bullet \text{Interaction}_{g2} \bullet \text{Interaction}_{g3} \bullet \text{Interaction}_{g4} \bullet STOP$$

IFD_n describes the nth interaction flow diagram, i.e. *Study_Algebra_Course* behavior, of the structural composition of the implementation view of *Kurdi University*. IFD_n is syntactically represented   as "$\text{Interaction}_{h1} \bullet \text{Interaction}_{h2} \bullet \text{Interaction}_{h3} \bullet \text{Interaction}_{h4} \bullet STOP$".

$$\text{IFD}_n =$$

$$\text{Interaction}_{h1} \bullet \text{Interaction}_{h2} \bullet \text{Interaction}_{h3} \bullet \text{Interaction}_{h4} \bullet STOP$$

Each interaction flow diagram may replicate itself a countably infinite time. Interaction_{xyz} stands for the yth replication of the zth interaction of the xth interaction flow diagram. $!\text{IFD}_m$ describes the infinite replication of the mth interaction flow diagram, i.e. *Study_Calculus_Course* behavior, of the structural composition of the implementation view of *Kurdi University*. $!\text{IFD}_m$ is   syntactically   represented   as "$(\text{Interaction}_{g11} \bullet \text{Interaction}_{g12} \bullet \text{Interaction}_{g13} \bullet \text{Interaction}_{g14} \bullet STOP)|$ $|(\text{Interaction}_{g21} \bullet \text{Interaction}_{g22} \bullet \text{Interaction}_{g23} \bullet \text{Interaction}_{g24} \bullet STOP)$ $||(\text{Interaction}_{g31} \bullet \text{Interaction}_{g32} \bullet \text{Interaction}_{g33} \bullet \text{Interaction}_{g34} \bullet STOP)...$ $....||(\text{Interaction}_{g\infty1} \bullet \text{Interaction}_{g\infty2} \bullet \text{Interaction}_{g\infty3} \bullet \text{Interaction}_{g\infty4} \bullet S TOP)$".

$!IFD_m =$

$(Interaction_{g11} \bullet Interaction_{g12} \bullet Interaction_{g13} \bullet Interaction_{g14} \bullet STOP)$
$\|$
$(Interaction_{g21} \bullet Interaction_{g22} \bullet Interaction_{g23} \bullet Interaction_{g24} \bullet STOP)$
$\|$
$(Interaction_{g31} \bullet Interaction_{g32} \bullet Interaction_{g33} \bullet Interaction_{g34} \bullet STOP)$
....
$\|$
$(Interaction_{g\infty1} \bullet Interaction_{g\infty2} \bullet Interaction_{g\infty3} \bullet Interaction_{g\infty4} \bullet STOP)$

$!IFD_n$ describes the infinite replication of the nth interaction flow diagram, i.e. *Study_Algebra_Course* behavior, of the structural composition of the implementation view of *Kurdi University*. $!IFD_n$ is syntactically represented as "$(Interaction_{h11} \bullet Interaction_{h12} \bullet Interaction_{h13} \bullet Interaction_{h14} \bullet STOP)\|$ $\|(Interaction_{h21} \bullet Interaction_{h22} \bullet Interaction_{h23} \bullet Interaction_{h24} \bullet STOP)$ $\|(Interaction_{h31} \bullet Interaction_{h32} \bullet Interaction_{h33} \bullet Interaction_{h34} \bullet STOP)$... ....$\|(Interaction_{h\infty1} \bullet Interaction_{h\infty2} \bullet Interaction_{h\infty3} \bullet Interaction_{h\infty4} \bullet S$ $TOP)$".

$!IFD_n =$

$(Interaction_{h11} \bullet Interaction_{h12} \bullet Interaction_{h13} \bullet Interaction_{h14} \bullet STOP)$
$\|$
$(Interaction_{h21} \bullet Interaction_{h22} \bullet Interaction_{h23} \bullet Interaction_{h24} \bullet STOP)$
$\|$
$(Interaction_{h31} \bullet Interaction_{h32} \bullet Interaction_{h33} \bullet Interaction_{h34} \bullet STOP)$
....
$\|$
$(Interaction_{h\infty1} \bullet Interaction_{h\infty2} \bullet Interaction_{h\infty3} \bullet Interaction_{h\infty4} \bullet STOP)$

Infinite-queue SBC process of the structural composition of the implementation view of *Kurdi University* is syntactically represented as $(!IFD_m)\|(!IFD_n)$ which equals to "(Interaction$_{g11}$•Interaction$_{g12}$•Interaction$_{g13}$•Interaction$_{g14}$•*STOP*)|

|(Interaction$_{g21}$•Interaction$_{g22}$•Interaction$_{g23}$•Interaction$_{g24}$•*STOP*)

‖(Interaction$_{g31}$•Interaction$_{g32}$•Interaction$_{g33}$•Interaction$_{g34}$•*STOP*)…

….‖(Interaction$_{g\infty1}$•Interaction$_{g\infty2}$•Interaction$_{g\infty3}$•Interaction$_{g\infty4}$•*S TOP*)‖(Interaction$_{h11}$•Interaction$_{h12}$•Interaction$_{h13}$•Interaction$_{h14}$•*STOP*)‖(Interaction$_{h21}$•Interaction$_{h22}$•Interaction$_{h23}$•Interaction$_{h24}$•*STOP*)‖(Interaction$_{h31}$•Interaction$_{h32}$•Interaction$_{h33}$•Interaction$_{h34}$•*STOP*).……‖(Interaction$_{h\infty1}$•Interaction$_{h\infty2}$•Interaction$_{h\infty3}$•Interaction$_{h\infty4}$•*STOP*)".

Kurdi University's Structural Composition of the Implementation View $\overset{\text{def}}{=\!=}$

(Interaction$_{g11}$•Interaction$_{g12}$•Interaction$_{g13}$•Interaction$_{g14}$•*STOP*)
‖
(Interaction$_{g21}$•Interaction$_{g22}$•Interaction$_{g23}$•Interaction$_{g24}$•*STOP*)
‖
(Interaction$_{g31}$•Interaction$_{g32}$•Interaction$_{g33}$•Interaction$_{g34}$•*STOP*) …
‖
(Interaction$_{g\infty1}$•Interaction$_{g\infty2}$•Interaction$_{g\infty3}$•Interaction$_{g\infty4}$•*STOP*)
‖
(Interaction$_{h11}$•Interaction$_{h12}$•Interaction$_{h13}$•Interaction$_{h14}$•*STOP*)
‖
(Interaction$_{h21}$•Interaction$_{h22}$•Interaction$_{h23}$•Interaction$_{h24}$•*STOP*)
‖
(Interaction$_{h31}$•Interaction$_{h32}$•Interaction$_{h33}$•Interaction$_{h34}$•*STOP*) …
‖
(Interaction$_{h\infty1}$•Interaction$_{h\infty2}$•Interaction$_{h\infty3}$•Interaction$_{h\infty4}$•*STOP*)

Observation Congruence of "the Design View" and "the Structural Composition of the Implementation View"

We syntactically represent infinite-queue SBC process P_{01} as $\text{IFD}_g\|\text{IFD}_h$ which equals to $(\text{Interaction}_{g1}\bullet\text{Interaction}_{g2}\bullet\text{Interaction}_{g3}\bullet\text{Interaction}_{g4}\bullet STOP)\|(\text{Interaction}_{h1}\bullet\text{Interaction}_{h2}\bullet\text{Interaction}_{h3}\bullet\text{Interaction}_{h4}\bullet STOP)$. We also syntactically represent infinite-queue SBC processes P_{02} as $(\text{Interaction}_{g2}\bullet\text{Interaction}_{g3}\bullet\text{Interaction}_{g4}\bullet STOP)\|(\text{Interaction}_{h1}\bullet\text{Interaction}_{h2}\bullet\text{Interaction}_{h3}\bullet\text{Interaction}_{h4}\bullet STOP)$, P_{03} as $(\text{Interaction}_{g3}\bullet\text{Interaction}_{g4}\bullet STOP)\|(\text{Interaction}_{h1}\bullet\text{Interaction}_{h2}\bullet\text{Interaction}_{h3}\bullet\text{Interaction}_{h4}\bullet STOP)$, P_{04} as $(\text{Interaction}_{g4}\bullet STOP\|\text{Interaction}_{h1}\bullet\text{Interaction}_{h2}\bullet\text{Interaction}_{h3}\bullet\text{Interaction}_{h4}\bullet STOP)$, P_{05} as $(\text{Interaction}_{h1}\bullet\text{Interaction}_{h2}\bullet\text{Interaction}_{h3}\bullet\text{Interaction}_{h4}\bullet STOP)$, P_{06} as $(\text{Interaction}_{g1}\bullet\text{Interaction}_{g2}\bullet\text{Interaction}_{g3}\bullet\text{Interaction}_{g4}\bullet STOP)\|(\text{Interaction}_{h2}\bullet\text{Interaction}_{h3}\bullet\text{Interaction}_{h4}\bullet STOP)$, P_{07} as $(\text{Interaction}_{g2}\bullet\text{Interaction}_{g3}\bullet\text{Interaction}_{g4}\bullet STOP)\|(\text{Interaction}_{h2}\bullet\text{Interaction}_{h3}\bullet\text{Interaction}_{h4}\bullet STOP)$, P_{08} as $(\text{Interaction}_{g3}\bullet\text{Interaction}_{g4}\bullet STOP)\|(\text{Interaction}_{h2}\bullet\text{Interaction}_{h3}\bullet\text{Interaction}_{h4}\bullet STOP)$, P_{09} as $(\text{Interaction}_{g4}\bullet STOP\|\text{Interaction}_{h2}\bullet\text{Interaction}_{h3}\bullet\text{Interaction}_{h4}\bullet STOP)$, P_{10} as $(\text{Interaction}_{h2}\bullet\text{Interaction}_{h3}\bullet\text{Interaction}_{h4}\bullet STOP)$, P_{11}

as

$(\text{Interaction}_{g1} \bullet \text{Interaction}_{g2} \bullet \text{Interaction}_{g3} \bullet \text{Interaction}_{g4} \bullet STOP) \| (\text{Interaction}_{h3} \bullet \text{Interaction}_{h4} \bullet STOP)$, P_{12} as $(\text{Interaction}_{g2} \bullet \text{Interaction}_{g3} \bullet \text{Interaction}_{g4} \bullet STOP) \| (\text{Interaction}_{h3} \bullet \text{Interaction}_{h4} \bullet STOP)$, P_{13} as $(\text{Interaction}_{g3} \bullet \text{Interaction}_{g4} \bullet STOP) \| (\text{Interaction}_{h3} \bullet \text{Interaction}_{h4} \bullet STOP)$, P_{14} as $(\text{Interaction}_{g4} \bullet STOP \| \text{Interaction}_{h3} \bullet \text{Interaction}_{h4} \bullet STOP)$, P_{15} as $(\text{Interaction}_{h3} \bullet \text{Interaction}_{h4} \bullet STOP)$, P_{16} as $(\text{Interaction}_{g1} \bullet \text{Interaction}_{g2} \bullet \text{Interaction}_{g3} \bullet \text{Interaction}_{g4} \bullet STOP \| (\text{Interaction}_{h4} \bullet STOP)$, P_{17} as $(\text{Interaction}_{g2} \bullet \text{Interaction}_{g3} \bullet \text{Interaction}_{g4} \bullet STOP \| (\text{Interaction}_{h4} \bullet STOP)$, P_{18} as $(\text{Interaction}_{g3} \bullet \text{Interaction}_{g4} \bullet STOP \| (\text{Interaction}_{h4} \bullet STOP)$, P_{19} as $(\text{Interaction}_{g4} \bullet STOP \| (\text{Interaction}_{h4} \bullet STOP)$, P_{20} as $(\text{Interaction}_{h4} \bullet STOP)$, P_{21} as $(\text{Interaction}_{g1} \bullet \text{Interaction}_{g2} \bullet \text{Interaction}_{g3} \bullet \text{Interaction}_{g4} \bullet STOP)$, P_{22} as $(\text{Interaction}_{g2} \bullet \text{Interaction}_{g3} \bullet \text{Interaction}_{g4} \bullet STOP)$, P_{23} as $(\text{Interaction}_{g3} \bullet \text{Interaction}_{g4} \bullet STOP)$, P_{24} as $(\text{Interaction}_{g4} \bullet STOP)$, P_{25} as $STOP$, respectively.

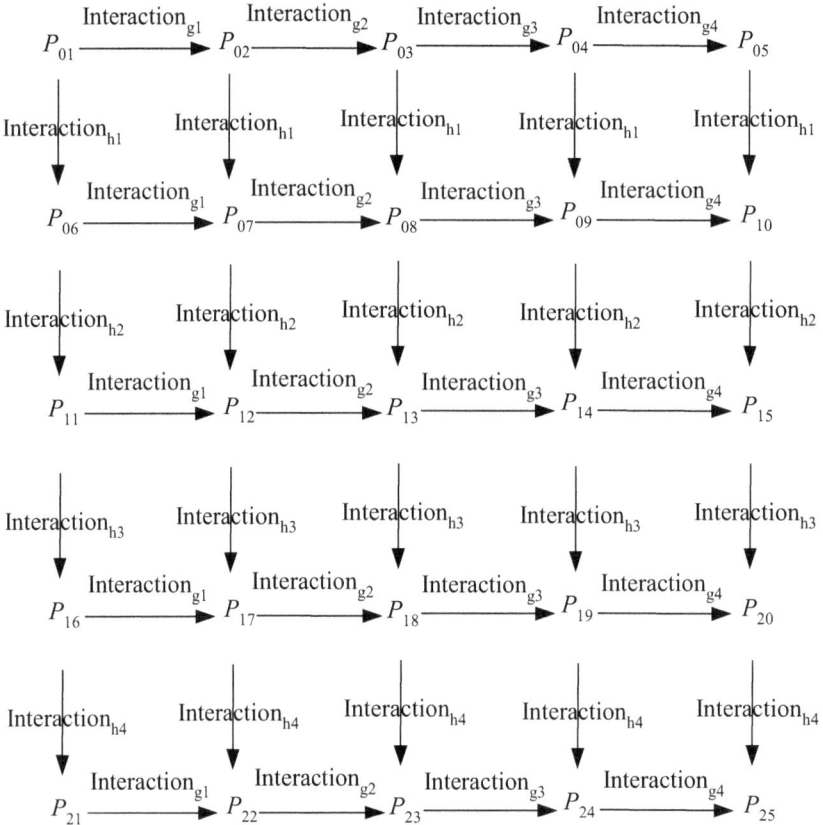

We syntactically represent infinite-queue SBC process Q_{01} as $\text{IFD}_m \| \text{IFD}_n$ which equals to $(\text{Interaction}_{g1} \bullet \text{Interaction}_{g2} \bullet \text{Interaction}_{g3} \bullet \text{Interaction}_{g4} \bullet STOP) \| (\text{Interaction}_{h1} \bullet \text{Interaction}_{h2} \bullet \text{Interaction}_{h3} \bullet \text{Interaction}_{h4} \bullet STOP)$. We also syntactically represent infinite-queue SBC processes Q_{02} as $(\text{Interaction}_{g2} \bullet \text{Interaction}_{g3} \bullet \text{Interaction}_{g4} \bullet STOP) \| (\text{Interaction}_{h1} \bullet \text{Interaction}_{h2} \bullet \text{Interaction}_{h3} \bullet \text{Interaction}_{h4} \bullet STOP)$, Q_{03} as $(\text{Interaction}_{g3} \bullet \text{Interaction}_{g4} \bullet STOP) \| (\text{Interaction}_{h1} \bullet \text{Interaction}_{h2} \bullet \text{Interaction}_{h3} \bullet \text{Interaction}_{h4} \bullet STOP)$, Q_{04} as $(\text{Interaction}_{g4} \bullet STOP \| \text{Interaction}_{h1} \bullet \text{Interaction}_{h2} \bullet \text{Interaction}_{h3} \bullet \text{Interaction}_{h4} \bullet STOP)$, Q_{05} as $(\text{Interaction}_{h1} \bullet \text{Interaction}_{h2} \bullet \text{Interaction}_{h3} \bullet \text{Interaction}_{h4} \bullet STOP)$, Q_{06} as $(\text{Interaction}_{g1} \bullet \text{Interaction}_{g2} \bullet \text{Interaction}_{g3} \bullet \text{Interaction}_{g4} \bullet STOP) \| (\text{Interaction}_{h2} \bullet \text{Interaction}_{h3} \bullet \text{Interaction}_{h4} \bullet STOP)$, Q_{07} as $(\text{Interaction}_{g2} \bullet \text{Interaction}_{g3} \bullet \text{Interaction}_{g4} \bullet STOP) \| (\text{Interaction}_{h2} \bullet \text{Interaction}_{h3} \bullet \text{Interaction}_{h4} \bullet STOP)$, Q_{08} as $(\text{Interaction}_{g3} \bullet \text{Interaction}_{g4} \bullet STOP) \| (\text{Interaction}_{h2} \bullet \text{Interaction}_{h3} \bullet \text{Interaction}_{h4} \bullet STOP)$, Q_{09} as $(\text{Interaction}_{g4} \bullet STOP \| \text{Interaction}_{h2} \bullet \text{Interaction}_{h3} \bullet \text{Interaction}_{h4} \bullet STOP)$, Q_{10} as $(\text{Interaction}_{h2} \bullet \text{Interaction}_{h3} \bullet \text{Interaction}_{h4} \bullet STOP)$, Q_{11} as $(\text{Interaction}_{g1} \bullet \text{Interaction}_{g2} \bullet \text{Interaction}_{g3} \bullet \text{Interaction}_{g4} \bullet STOP) \| (\text{Interaction}_{h3} \bullet \text{Interaction}_{h4} \bullet STOP)$, Q_{12} as $(\text{Interaction}_{g2} \bullet \text{Interaction}_{g3} \bullet \text{Interaction}_{g4} \bullet STOP) \| (\text{Interaction}_{h3} \bullet \text{Interaction}_{h4} \bullet STOP)$, Q_{13} as $(\text{Interaction}_{g3} \bullet \text{Interaction}_{g4} \bullet STOP) \| (\text{Interaction}_{h3} \bullet \text{Interaction}_{h4} \bullet STOP)$, Q_{14} as $(\text{Interaction}_{g4} \bullet STOP \| \text{Interaction}_{h3} \bullet \text{Interaction}_{h4} \bullet STOP)$, Q_{15} as

$(\text{Interaction}_{h3} \bullet \text{Interaction}_{h4} \bullet STOP)$, Q_{16} as $(\text{Interaction}_{g1} \bullet \text{Interaction}_{g2} \bullet \text{Interaction}_{g3} \bullet \text{Interaction}_{g4} \bullet STOP \| (\text{Interaction}_{h4} \bullet STOP)$, Q_{17} as $(\text{Interaction}_{g2} \bullet \text{Interaction}_{g3} \bullet \text{Interaction}_{g4} \bullet STOP \| (\text{Interaction}_{h4} \bullet STOP)$, Q_{18} as $(\text{Interaction}_{g3} \bullet \text{Interaction}_{g4} \bullet STOP \| (\text{Interaction}_{h4} \bullet STOP)$, Q_{19} as $(\text{Interaction}_{g4} \bullet STOP \| (\text{Interaction}_{h4} \bullet STOP)$, Q_{20} as $(\text{Interaction}_{h4} \bullet STOP)$, Q_{21} as $(\text{Interaction}_{g1} \bullet \text{Interaction}_{g2} \bullet \text{Interaction}_{g3} \bullet \text{Interaction}_{g4} \bullet STOP)$, Q_{22} as $(\text{Interaction}_{g2} \bullet \text{Interaction}_{g3} \bullet \text{Interaction}_{g4} \bullet STOP)$, Q_{23} as $(\text{Interaction}_{g3} \bullet \text{Interaction}_{g4} \bullet STOP)$, Q_{24} as $(\text{Interaction}_{g4} \bullet STOP)$, Q_{25} as $STOP$, respectively.

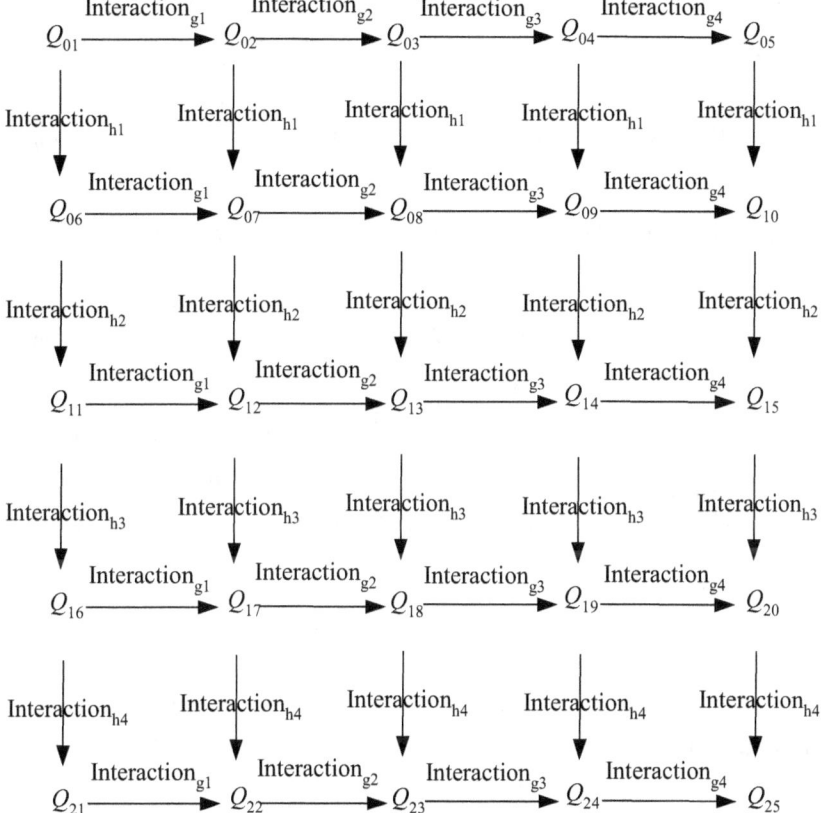

We can easily verify that $S = \{(P_{01}, Q_{01}), (P_{02}, Q_{02}), (P_{03}, Q_{03}), (P_{04}, Q_{04}), (P_{05}, Q_{05}), (P_{06}, Q_{06}), (P_{07}, Q_{07}), (P_{08}, Q_{08}), (P_{09}, Q_{09}), (P_{10}, Q_{10}), (P_{11}, Q_{11}), (P_{12}, Q_{12}), (P_{13}, Q_{13}), (P_{14}, Q_{14}), (P_{15}, Q_{15}), (P_{16}, Q_{16}), (P_{17}, Q_{17}), (P_{18}, Q_{18}), (P_{19}, Q_{19}), (P_{20}, Q_{20}), (P_{21}, Q_{21}), (P_{22}, Q_{22}), (P_{23}, Q_{23}), (P_{24}, Q_{24}), (P_{25}, Q_{25})\}$ is a bisimulation.

Using the *S* bisimulation, we then are able to verify that P_{01} and Q_{01} are observation congruent because (1) $P_{01} \xrightarrow{\text{Interaction}_{g1}} P_{02}$, then we have Q_{02} that $Q_{01} \overset{\text{Interaction}_{g1}}{\Longrightarrow} Q_{02}$ and $P_{02} \overset{\sim}{\approx} Q_{02}$, and (2) $P_{01} \xrightarrow{\text{Interaction}_{h1}} P_{06}$, then we have Q_{06} that $Q_{01} \overset{\text{Interaction}_{h1}}{\Longrightarrow} Q_{06}$ and $P_{06} \overset{\sim}{\approx} Q_{06}$, and (3) $Q_{01} \xrightarrow{\text{Interaction}_{g1}} Q_{02}$, then we have P_{02} that $P_{01} \overset{\text{Interaction}_{g1}}{\Longrightarrow} P_{02}$ and $P_{02} \overset{\sim}{\approx} Q_{02}$, and (4) $Q_{01} \xrightarrow{\text{Interaction}_{h1}} Q_{06}$, then we have P_{06} that $P_{01} \overset{\text{Interaction}_{h1}}{\Longrightarrow} P_{06}$ and $P_{06} \overset{\sim}{\approx} Q_{06}$.

$P_{01} = Q_{01}$ means $\text{IFD}_g \| \text{IFD}_h = \text{IFD}_m \| \text{IFD}_n$, and $\text{IFD}_g \| \text{IFD}_h = \text{IFD}_m \| \text{IFD}_n$ means $!(\text{IFD}_g \| \text{IFD}_h) = !(\text{IFD}_m \| \text{IFD}_n)$, and $!(\text{IFD}_g \| \text{IFD}_h) = !(\text{IFD}_m \| \text{IFD}_n)$ means $(!\text{IFD}_g) \| (!\text{IFD}_h) = (!\text{IFD}_m) \| (!\text{IFD}_n)$. So, there is observation congruence of "the design view of *Kurdi University*" and "the structural composition of the implementation view of *Kurdi University*".

Conclusively, the implementation view of *Kurdi University* is one level down structural decomposition (with observation congruence verification) of the design view of *Kurdi University*.